The Antarctic Voyage of HMAS *Wyatt Earp*

The Antarctic Voyage of HMAS *Wyatt Earp*

Phillip Law

Allen & Unwin

First published in 1995

Allen & Unwin Pty Ltd
9 Atchison Street, St Leonards, NSW 2065 Australia

National Library of Australia
Cataloguing-in-Publication entry:

Law, P.G. (Phillip Garth), 1912– .
The Antarctic Voyage of HMAS Wyatt Earp

Includes index.
ISBN 1 86373 803 7.

1. Wyatt Earp (Ship). 2. Australian National Antarctic Research Expeditions. 3. Scientific expeditions – Antarctica. 4. Australian Antarctic Territory (Antarctic Regions) – Discovery and exploration. I. Title.

919.8904

Set in Palatino and 11/13pt Bembo
by Graphicraft Typesetters Ltd, Hong Kong
Printed by South Wind Productions, Singapore

10 9 8 7 6 5 4 3 2 1

Frontispiece: HMAS *Wyatt Earp* being moored to an ice floe in Antarctica, 1948 (Photo by L. Le Guay)

Contents

Acknowledgements

I PARTICULARLY APPRECIATE the support of the Australian Antarctic Foundation, without whose financial assistance this book would not have been published.

I wish to thank Nancy Patton for kindly volunteering her time to type the first draft of this book, and Julie Di Pietro, my part-time secretary, for her work on the later drafts.

The Department of Defence (Navy Office) was helpful in providing historical information about the *Wyatt Earp* and details of the ship's company, while the Australian War Memorial, Canberra, assisted with photographs. Norman Tame provided me with the section layout sketch of the ship.

Note on Measurements

THROUGHOUT THE TEXT of this book, the non-metric measurements of the late 1940s are used. Those who wish to convert to the metric system now used will find the following helpful:

1 yard	= 0.9144 metres
1 mile	= 1.609 kilometres
1 pint	= 0.568 litres
1 quart	= 1.137 litres
1 ton (short)	= 0.907 tonne
1 ton (long)	= 1.016 tonnes
1 ounce	= 28.350 grams
2.471 acres	= 1 hectare
1.308 cubic yards	= 1 cubic metre

ANARE 1947-48 The voyage of HMAS *Wyatt Earp*

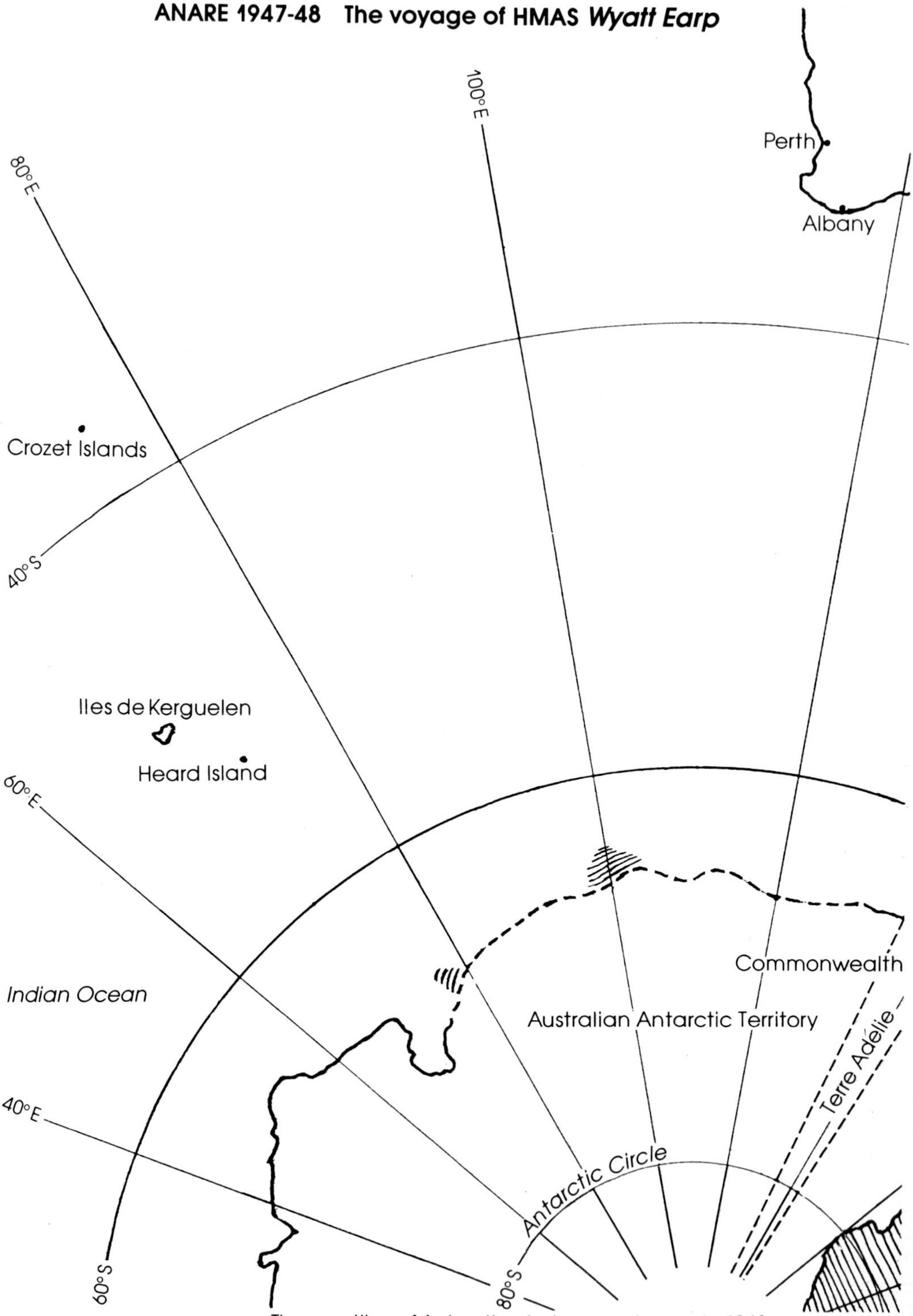

The coastline of Antarctica is shown as known in 1948

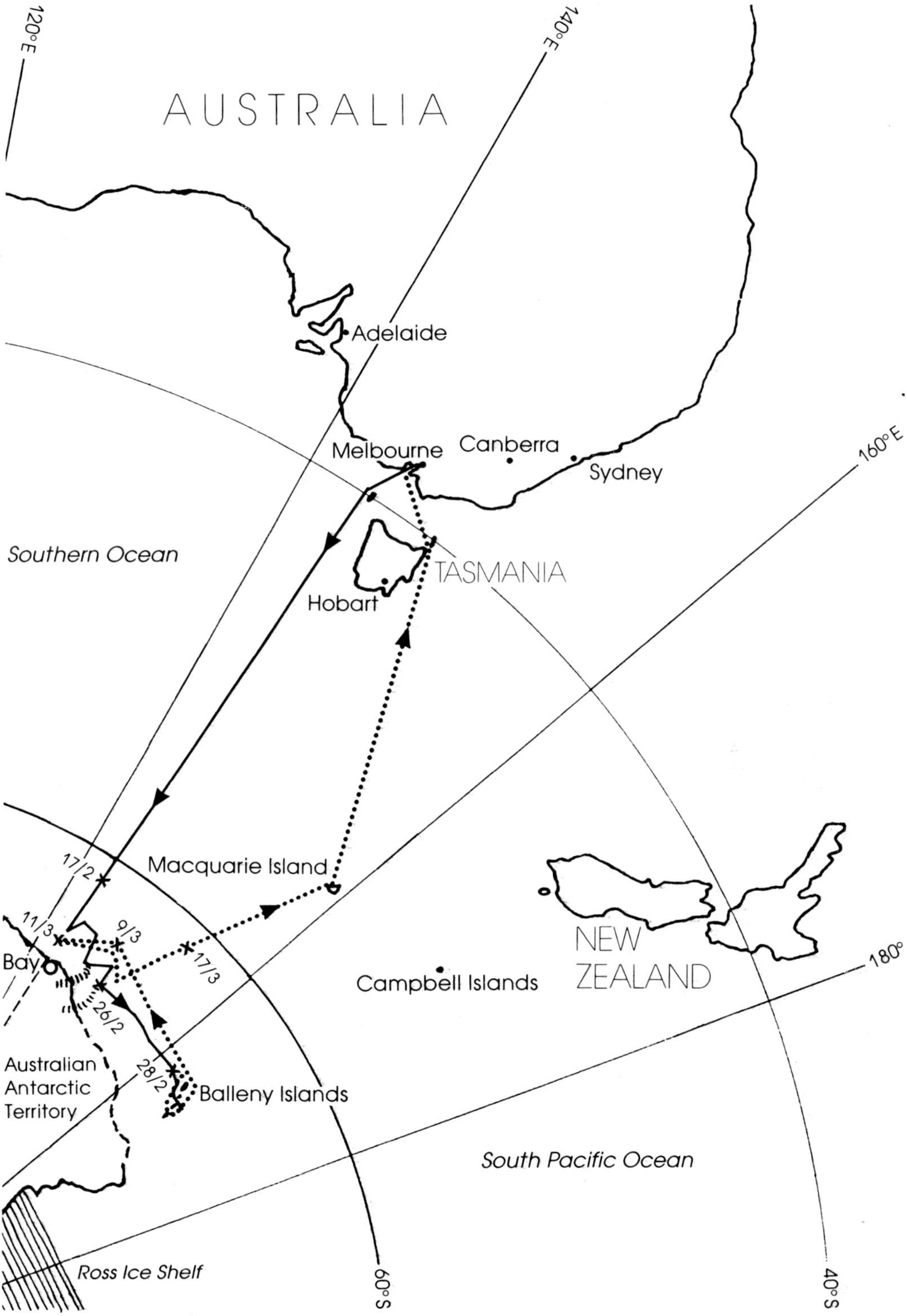
120°E
140°E
AUSTRALIA
Adelaide
Melbourne
Canberra
Sydney
160°E
Southern Ocean
TASMANIA
Hobart
Macquarie Island
17/2
11/3
9/3
17/3
Bay
26/2
28/2
NEW
ZEALAND
Campbell Islands
180°
Australian
Antarctic
Territory
Balleny Islands
South Pacific Ocean
Ross Ice Shelf
60°S
40°S

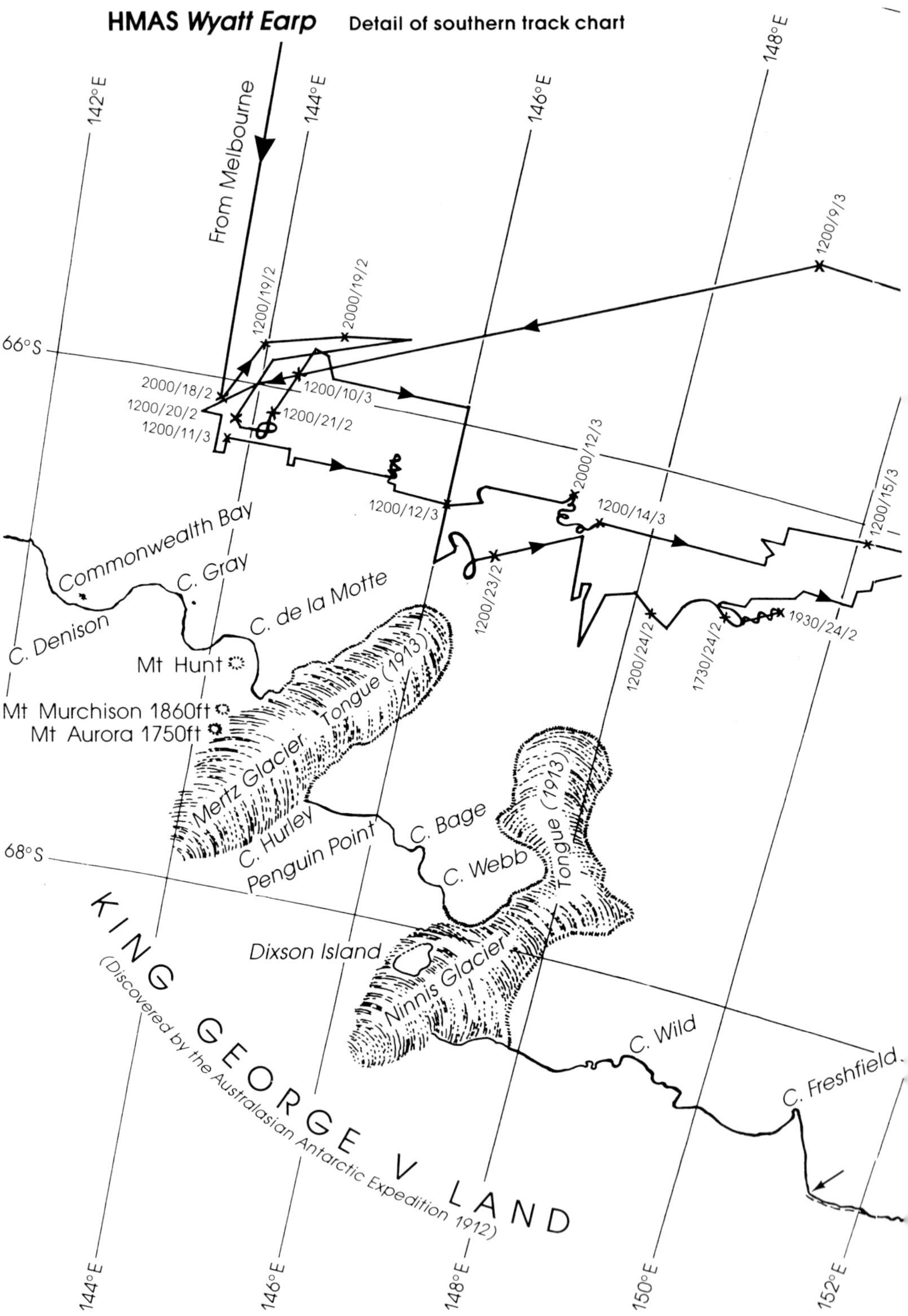
HMAS Wyatt Earp
Detail of southern track chart
From Melbourne
142°E
144°E
146°E
148°E
150°E
152°E
66°S
68°S
1200/19/2
2000/19/2
1200/9/3
2000/18/2
1200/20/2
1200/11/3
1200/10/3
1200/21/2
1200/12/3
2000/12/3
1200/14/3
1200/15/3
1200/23/2
1200/24/2
1730/24/2
1930/24/2
Commonwealth Bay
C. Gray
C. Denison
C. de la Motte
Mt Hunt
Mt Murchison 1860ft
Mt Aurora 1750ft
Mertz Glacier Tongue (1913)
C. Hurley
Penguin Point
C. Bage
C. Webb
Ninnis Glacier Tongue (1913)
Dixson Island
C. Wild
C. Freshfield
KING GEORGE V LAND
(Discovered by the Australasian Antarctic Expedition 1912)

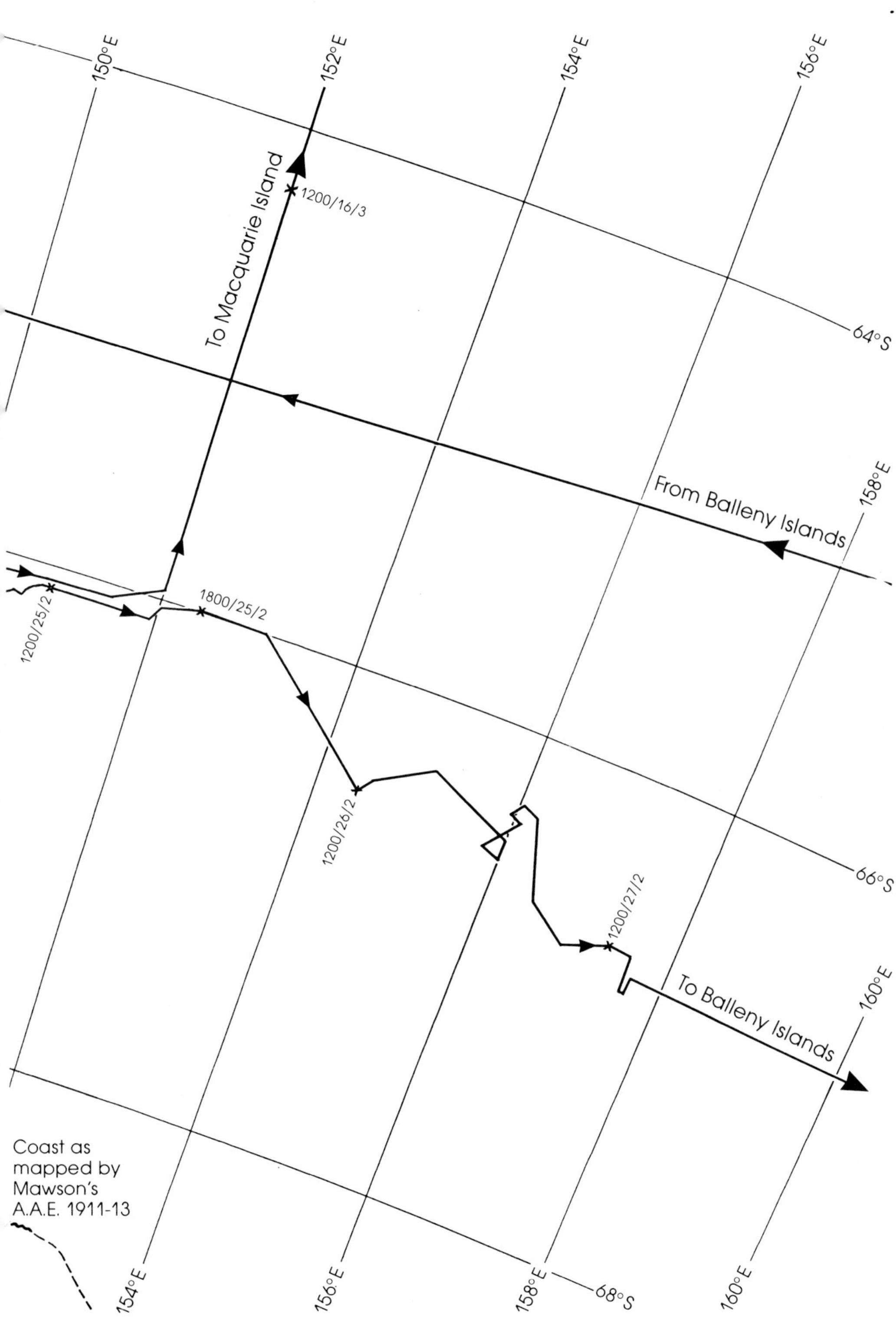
150°E
152°E
154°E
156°E
158°E
160°E
64°S
66°S
68°S
To Macquarie Island
1200/16/3
From Balleny Islands
1200/25/2
1800/25/2
1200/26/2
1200/27/2
To Balleny Islands
Coast as mapped by Mawson's A.A.E. 1911-13

1

The ANARE Program Gets Under Way

AUSTRALIAN INTEREST IN Antarctica goes back some 150 years. The earliest ventures were those of sealers and whalers, who were motivated by visions of quick profits, rather than by any sense of curiosity. Among the pioneers were Frederick Hasselburgh, who discovered Macquarie Island in 1810, Richard Siddons and D. Taylor, who took sealing expeditions in 1820 from Sydney to the South Shetland Islands, and Samuel Harvey, who in 1831 pushed south from Hobart to reach 72° south latitude, in what was later to be called the Ross Sea.

During the nineteenth century, a number of famous Antarctic voyagers of various nationalities called at Sydney, Melbourne or Hobart on their way south. They included Bellingshausen (1819–21), Balleny (1838–39), Dumont d'Urville (1837–40), Wilkes (1838–42), Ross (1839–43) and Nares (1872–76). A sealing venture from Hobart, led by Captain Robertson in the *Offley*, operated at Heard Island in 1858.

The first Australian Antarctic Exploration Committee was set up by the Royal Society of Victoria in Melbourne in 1886. It worked for twenty years trying to stimulate interest in an Antarctic expedition, and to raise the necessary funds. It was unsuccessful.

It was left to a Melbourne resident of Norwegian nationality, Henrik Bull, to solve the financial problem. He persuaded the Norwegian whaling magnate, Svend Foyn, to sponsor a purely whaling expedition to the Antarctic in 1894. A young Norwegian immigrant who had been in Australia since 1888, Carstens Borchgrevink, applied to Bull for a position as scientific observer. Bull took him on as a general hand but, despite this, Borchgrevink devoted a lot of his time on the expedition to scientific observations.

Bull's ship, the *Antarctic*, flying the Norwegian flag, reached the Antarctic continent. Bull landed briefly with a party at Cape Adare, at the entrance to the Ross Sea, in January 1895.

Borchgrevink, his enthusiasm fired by his Antarctic experience, succeeded in organising a British expedition in the ship *Southern Cross*, with the sponsorship of a publisher, Sir George Newnes. On its way south the ship called at Melbourne, where it took on a young Tasmanian physicist, Louis Bernacci, who had been working on the staff of the Melbourne Observatory. He was to become the first Australian-born person to land and winter on the Antarctic continent.

The Borchgrevink expedition set up and occupied a station at Cape Adare in 1899. Its remnants can still be seen there today. They made valuable observations in zoology, geology, meteorology and terrestrial magnetism. However, the steep terrain behind their base denied them access to the Antarctic hinterland.

The years 1900–14 saw great progress in Antarctic exploration, through the famous expeditions of Scott, Amundsen, Shackleton and Mawson. To each of these, except Amundsen's, Australia made important contributions.

Scott's *Discovery* expedition of 1901–3 obtained a grant of 250 pounds from the Australian Antarctic Exploration Committee and 1,000 pounds from the Queensland government. Scott enlisted Louis Bernacci as physicist.

When Shackleton brought the British Antarctic Expedition to Melbourne in the *Nimrod* in 1907, he was in debt and in danger from bailiffs. Professor Edgeworth David of the University of Sydney persuaded the Australian government to donate 50,000 pounds to the expedition. When the New Zealand government contributed 1,000 pounds, Shackleton's main financial worries were over.

Shackleton enlisted Douglas Mawson as physicist and B. Armstrong of Melbourne as a general hand. Professor David and one of his geological staff, Leo Cotton, signed up for the round trip on the *Nimrod*. When they arrived at McMurdo, however, Shackleton persuaded David to remain for the whole year as his Chief Scientist. The first officer of the *Nimrod*, John King Davis, later became master of this ship and of Mawson's *Aurora* and *Discovery*.

David and Mawson made notable contributions in geology and geophysics to the achievements of Shackleton's expedition. A six-man party led by David and including Mawson made the first ascent of the Antarctic volcano, Mt Erebus (altitude 12,450 feet). Later, a three-man party of David, Mawson and MacKay manhauled a sledge to the South Magnetic Pole, a return journey of 1,260 miles, in 122 days. This was a major achievement and one that has never been adequately recognised.

Scott's second expedition, in 1911–13 in the *Terra Nova*, also obtained a grant from the Australian government, this time of 2,500 pounds. Two Australian scientists accompanied Scott, Griffith Taylor and Frank Debenham. Taylor later established the first Department of Geography in Australia, at the University of Sydney, and Debenham became the foundation Director of the Scott Polar Research Institute in Cambridge.

After returning from Shackleton's expedition, Mawson set about organising one of his own. He obtained financial assistance from numerous private sources in both Australia and Britain, and donations over a two-year period from the Commonwealth government (5,000 pounds), the New South Wales government (7,000 pounds), the Victorian government (600 pounds), and the Australian Association for the Advancement of Science (1,000 pounds). His Australasian Antarctic Expedition (AAE) of 1911–13 established three stations: at Macquarie Island, at Commonwealth Bay (George V Land) and on the Shackleton Ice Shelf (Queen Mary Land).

In February 1912, the *Aurora*, commanded by Captain J.K. Davis, sailed along off the coast of Antarctica between 140° east longitude and 90° east longitude, naming Wilkes Land, Queen Mary Land, the Davis Sea and the Shackleton Ice Shelf. Sledging parties from the Western Base mapped the coast of Queen Mary Land, while others from Commonwealth Bay explored 150 miles west, 300 miles south, and 300 miles east of their base. The scientific reports of the expedition fill twenty-two volumes.

Shackleton's Imperial Trans-Antarctic Expedition of 1914–17 also included Australians. He took with him Mawson's photographer, Frank Hurley, and his Ross Sea party included Australians R.W. Richards, A.K. Jack, and C.O. Gaze. Richards was the hero of that party, and was awarded the George Cross.

The Australian adventurer, Sir Hubert Wilkins, became famous because of his pioneer flight over the North Pole. In 1921 he joined Cope's British four-man expedition to Graham Land, and in 1922 he was with Shackleton's last expedition. He and pilot Eielson made the first flight in the Antarctic, in November 1928—1,300 miles over Graham Land. He later accompanied the American explorer, Lincoln Ellsworth, on his 1935–36 and 1939 expeditions.

When Mawson returned from his Australasian Antarctic Expedition, the outbreak of war in 1914 precluded any further Australian activity in the region for some years. In the 1920s, however, Mawson led a resurgence of national interest in Antarctic exploration. Following discussions between the British and Australian governments, it was agreed that Britain would make available the ship *Discovery* for an expedition to be led by Mawson in the summer of 1929–30. This would, it was hoped, strengthen any claim the British might make to Antarctic territory in the so-called Australian sector. Thus arose the British–Australian–New Zealand Antarctic Expedition (BANZARE), a private expedition supported by the three governments concerned, with private donations.

Two voyages were made in the summers of 1929–30 and 1930–31. The *Discovery* visited Macquarie Island, Iles de Kerguelen and Heard Island in the sub-Antarctic and cruised along the coast of Antarctica between Terre Adélie in the east and Enderby Land in the west. A landing was made at Proclamation Island (Enderby Land), where Mawson read a statement laying claim on behalf of King George V of Britain to all lands between longitudes 47° east and 173° east, and south of latitude 65°S.

Other landings were made at Cape Bruce, Kemp Land (where a copy of the Proclamation was deposited beneath a cairn), and at Scullin Monolith, MacRobertson Land. Several flights were made in a Moth aeroplane, discovering from the air the BANZARE coast and Princess Elizabeth Land, and confirming the existence of the Knox coast, much further south than Wilkes had described it in the nineteenth century.

Another Australian, John Rymill of Adelaide, made important contributions to Antarctic knowledge, even though the expedition he led was a British one. This was the Graham Land expedition of

1934–37, which carried out high quality scientific work and cartography in the Antarctic Peninsula region.

World War II again caused a hiatus in Australian Antarctic work. When it was resumed with the 1947 ANARE, it was of a completely different nature. In place of the privately organised, short-term 'amateur' expeditions of the pre-War era, there developed government financed, highly professional activities, leading up to the great multinational attack on Antarctica during the International Geophysical year of 1957–58.

Australia's claim to sovereignty over Australian Antarctic territory was based, in the first instance, upon the work of Sir Douglas Mawson's two expeditions (the AAE of 1913–15 and the BANZARE of 1929–31), and to a lesser extent upon early British work by Biscoe (1830–32), Kemp (1833–34) and Balleny (1838–39).

A British Order-in-Council dated 7 February 1933 claimed sovereign rights over Antarctic territory, other than Terre Adélie (French), between longitudes 160°E and 45°E and south of latitude 60°S. It placed the region under the control of the Commonwealth of Australia from the date on which the necessary legislation was passed in the Australian Parliament. The Australian Antarctic Territory Acceptance Act received royal assent on 30 June 1933, and was proclaimed in August 1936.

Following Mawson's BANZARE, 1929–31, further Australian work in the Antarctic was precluded, first by the Great Depression and, later, by World War II. But Mawson had not abandoned his idea of a permanent Australian station on the Antarctic continent. During 1946, he waged a campaign to persuade the Commonwealth government to initiate immediate measures to support and consolidate Australia's territorial claims in Antarctica, which were under threat because of the activities of two nations, the Norwegians, who made five summer cruises along the coast of Greater Antarctica between 1929 and 1937, and the Americans, who planned a massive naval circumnavigation of Antarctica for the summer of 1946–47.

In response to Mawson's urgings, Dr Herbert Evatt, Minister for External Affairs, convened an inter-Departmental committee meeting on 2 December 1946. This was attended by Mawson and a representative of the Council for Scientific and Industrial Research. Mawson pointed out that most of Australian Antarctic Territory was still unexplored. He also stressed that any expedition initiated for political reasons should carry out as much scientific research as its logistical limitations would allow. The Antarctic, he said, was an almost virgin field for investigation in a number of scientific disciplines. He proposed that any expeditionary organisation established for such tasks should re-establish the meteorological observatory on Macquarie Island he had set up in 1911 and that had not been continued beyond 1915.

The committee recommended that the departments concerned should develop concrete plans for an expedition to Australian Antarctic Territory, using a naval ship equipped with a suitable aircraft. The object was to find an ice-free area on the continent that could be used as the site for a permanent base. Mawson suggested that the ship HMAS *Wyatt Earp*, owned by the Navy but lying uncommissioned in the Torrens River at Port Adelaide, be refitted for use by the expedition. Cabinet accepted this recommendation on 20 December 1946, and proposed that an executive committee be formed to develop plans and estimates for an expedition in the summer of 1947–48, that Mawson should act as adviser to this committee, and that an executive officer be appointed to organise the journey.

The first meeting of the Executive Committee on Exploration and Exploitation was held on 4 January 1947, chaired by the Permanent Head of the Department of External Affairs, Dr John Burton. Thereafter events moved swiftly. With various personnel, and under several different titles, a number of committee meetings was held. The following proposals were finally put forward to the Department of External Affairs:

- that an Executive Planning Committee be formed under the Department;
- that preliminary plans for an expedition be drawn up at an estimated cost of 250,000 pounds;
- that the ship *Wyatt Earp* be refitted by the Navy;
- that an LST, a Landing Ship, Tanks, be provided by the Navy to establish a scientific station at Macquarie Island;
- that Group Captain Stuart Campbell (RAAF retd) be seconded from his position as Director of Air Navigation and Safety in the Department of Civil Aviation, to act as Chief Executive officer of the expedition. (Campbell had been senior aircraft pilot on Mawson's two BANZARE cruises).

Campbell was duly appointed, and took up duty on 12 May. He was supported by a temporary Secretary, Mr A. Cunningham

Group Captain Stuart Campbell (RAAF retd), 1947.

Tweedie. At a meeting of the Executive Planning Committee held on 26 May, it was recommended that a scientific officer be appointed to the expedition to assist the Chief Executive Officer.

The British government had been advised by the Australians of the moves being made. It asked that any expedition plans should include the occupation of Heard Island. The reason for this request was that, although the island had first been sighted by an English sealer, Captain Peter Kemp, in 1833 and later by an American whaler, Thomas Long, in 1849, these sightings had not been published. The credit for discovering the island therefore went to Captain John Heard, a merchant captain on a 'great circle' voyage from Boston to Melbourne in 1853. The first landing was made by another American sealer, Captain Erasmus Darwin Rogers in 1855. In 1946, rumours had begun to circulate that the Americans were planning to despatch an expedition to occupy the island, which had never been officially claimed by any nation.

The British request was classified as 'Secret' and Australian plans to include Heard Island in the ANARE program were kept very quiet in the early stages.

The outcome of the deliberations was that the Australian government approved recommendations for scientific stations to be established both at Heard Island and Macquarie Island; proposed that they be maintained for at least five years; and that a reconnaissance of King George V Land should be carried out to seek a suitable site for a permanent Antarctic station. It approved the use of two naval ships, the *Wyatt Earp*, captained by Commander Carl Oom, RAN, for the Antarctic voyage, and LST 3501, captained by Lieutenant-Commander George Dixon, RANVR, for the two island voyages. The Council for Scientific and Industrial Research (CSIR) was to accept responsibility for the financial aspects and the Royal Australian Air Force was to act as purchasing agent. Victualling would be the responsibility of the Royal Australian Navy. The scientific plans were to include meteorology, geology, biology, glaciology, geomagnetism, cosmic ray investigations and oceanography. Cartographic work associated with any exploration done would be carried out by the National Mapping Division of the Department of the Interior, meteorology by the Commonwealth Bureau of Meteorology, and geomagnetism and geology by the Commonwealth Bureau of Mineral Resources. Mr Trevor Heath was to become the full-time Secretary in place of Tweedie. He took up his appointment in June.

The way in which I became associated with the Australian National Antarctic Research Expedition, or ANARE, was unusual. Early in my boyhood, I had become interested in camping, bushwalking and mountaineering; later, this extended naturally to skiing in the Australian Alps. As a result of my interest in snow and ice, I began to explore Antarctic literature and soon had read most of the narratives of the classical expeditions.

By 1947, I was a Lecturer in the Physics Department at the University of Melbourne. My Professor was Leslie Martin, later to become Sir Leslie and Chairman of the Australian Universities Commission. In 1947 he was Scientific Adviser on defence matters to the federal government, and made frequent trips to Canberra. One day in early June that year, I was walking along a corridor in the Physics Department with him when he said, out of the blue: 'I have just been to Canberra. They are having difficulty in finding a Chief Scientist for the new Antarctic expedition.'

I had heard vague rumours of such an expedition, but had not been able to find out anything specific. I had wondered how I might obtain a job on it, and had considered writing to Sir Douglas Mawson to enquire. Now here was my Professor actually involved with it.

I said, immediately: 'Did you mention my name?'

'Surely you wouldn't be interested in going on that, would you?' he asked.

'I would give my right arm to go,' I said.

'Goodness me,' he replied. 'I shall go and ring up!'

This he did and, in a very short time, I was asked to present myself for an interview. This resulted in my appointment, from 1 August, as Senior Scientific Officer of the expedition, on secondment for twelve months from the University of Melbourne. My job was to co-ordinate the development of a number of scientific programs proposed by interested government departments and universities.

Before my actual starting date, a July meeting of the Executive Planning Committee was scheduled, and I was invited to attend. I had the pleasure of being present, on 9 July, at the meeting at which a name for the new expedition was determined: the Australian National Antarctic Research Expedition, or ANARE. I was not to know then that the ANARE would be the central focus of most of my activities for the rest of my life!

On 1 August 1947, I presented myself at expedition headquarters, a suite of rooms made available by the Department of the Navy in the Victoria Barracks, St Kilda Road, Melbourne. I was welcomed by Stuart Campbell, introduced to Secretary Trevor Heath, a middle-aged Oxford graduate, and to several female typists. They were the total establishment in the city. Out at Tottenham, a lone storeman, George Smith, sat in an empty shed at the No. 1 Stores Depot of the RAAF, waiting for supplies to begin rolling in. He had been appointed on 30 June.

The environment was, to say the least, dismal. The barracks presented to St Kilda Road a high bluestone wall more akin to a gaol than to an administrative headquarters. I passed through a guarded gate, walked along an asphalted court between lines of severe, two-storied, bluestone buildings, climbed a flight of wooden stairs and then, at the end of a narrow lino-floored corridor, entered the small, cell-like rooms allocated for our use. Each was similar. There were two small windows looking out on to the asphalt below and the bluestone blocks opposite, a brown linoleum floor, two tables and several chairs. A couple of rooms boasted filing cabinets. Walls were white plaster and the doors, floors and blinds were brown. I sat down at a table in an otherwise empty room and reflected upon the task ahead of me.

Phillip Law, Senior Scientific Officer ANARE, at Albert Park Barracks, 1947.

Early in 1947 Professor Martin had invited me to join a research team in the Physics Department that was to develop a program of research in cosmic rays. When the Antarctic expedition got under way, Martin proposed that the program should include cosmic ray studies. He suggested that measurements be made at Heard Island and Macquarie Island, to study long-term variations in the composition and intensity of the rays, and also from on board the *Wyatt Earp*, to measure latitude variations between Australia and Antarctica. By the time I took up my duties at the Victoria Barracks I was already well involved with a team of physicists working to build equipment for these tasks.

As a physicist I was familiar with various aspects of geophysics: meteorology, gravitation, seismology, geomagnetism, upper atmosphere physics and cosmic rays. It was obvious that Heard Island and Macquarie Island would provide opportunities for unique observatory studies in these fields. Even more obvious was the need for investigation of the zoological and botanical species and the general ecology of these sub-Antarctic islands and of their geology.

The question of 'What science?', then, was not a difficult one. Preliminary suggestions for possible areas of scientific research had already been presented in papers tabled at a meeting chaired by Dr F.W.G. White, the head of CSIR, on 6 March 1947. This had been attended by the heads of a number of scientific Commonwealth Departments, and by Sir Douglas Mawson.

The problems presented were how to gain the support of the responsible authorities, how to stimulate immediate action, how to appoint the expedition scientists and how to provide their apparatus and supporting equipment. Everything had to be ready by mid-December.

My main contacts in co-ordinating the scientific program were N.H. Warren, Director of the Commonwealth Meteorological Bureau, Jack Rayner, Deputy-Director of the Bureau of Geology, Geophysics and Mineral Resources, Professor Leslie Martin, Physics Department, University of Melbourne, Professor W.J. Dakin, Zoology Department, University of Sydney, Dr Frederick White, Chairman, CSIR and Dr George Munro, Radiophysics Laboratory, CSIR. These departments knew what they would like to do; it was a matter of listing the proposed projects, deciding which could be included given the logistical limitations of the ship transportation and the island stations' technical support systems, and then assessing the probable costs.

I prepared details of the scientific programs for the three expeditions. These were duly approved by the Executive Planning Committee.

The program for the *Wyatt Earp* voyage was ambitious:

1. To attempt to reach the coast of George V Land somewhere between Cape Freshfield and Commonwealth Bay, and to discover, if possible, a suitable site for a future Australian Antarctic station.
2. To carry out meteorological observations throughout the voyage (synoptic observations every six hours, air temperature, pressure and relative humidity, using instruments mounted in a Stevenson Screen in free-moving air on the after boat deck; daily measurements of wind speeds and directions at various heights using pilot balloons launched from the deck and tracked by shipboard theodolite).
3. To make geomagnetic observations of the Earth's magnetic field at any points where landings could be made in the Antarctic.

Their Husbands Sailing Soon for Polar Seas

Interviews By ELVA BREEN

Mrs. F. Loewe — Mrs. T. Bond — Mrs. P. G. Law

WHEN the Australian expedition leaves soon for the Antarctic, a group of men will exchange warmth and home-cooked meals for bleak cold and a diet of tinned foods, penguin and seal meat. Family men, most of them, they will leave wives whose only link with the polar region will be radioed messages from their husbands.

Dr. Fritz Loewe, his two daughters and the family cat were interested listeners when I talked with Mrs. Loewe, whose husband will be meteorologist, geographer and oceanographer on the trip.

Past experience has not endeared polar exploration to Mrs. Loewe. Her husband has made three other journeys to polar regions. On an expedition to Greenland in 1930 he was isolated on an ice shelf for five months. His toes were amputated as a result of frostbite.

Mrs. Loewe accompanied her husband on his third trip to Greenland in 1932 — an expedition to shoot backgrounds for the German film, SOS Iceberg. She and her husband still speak with a faint accent, but not so their daughters, Susanne, 13, and Ruth, 15, both of whom are pupils at the University High School.

Tall, slim and attractive Mrs. P. G. Law, wife of the senior science officer of the expedition, is an enthusiastic and able painter of portraits and landscapes. We talked in the lounge room of her South Yarra flat, in which her paintings and her husband's maps of the Antarctic jockey for wall space.

The Laws, who have been married for five years, met when both were University students. Mrs. Law is taking a keen interest in her husband's trip. She intends to "hold it over his head" when she proposes a visit to Europe in two years.

She has read everything she can find on the Antarctic. Books with such titles as Snow on the Hills and Home of the Blizzard fill her shelves.

Only two aspects of the expedition are giving her concern. First, that her husband might return with a beard, and secondly, that he might not have hot water for showers on board the Wyatt Earp. To be on the safe side, she has included an electric jug in his kit!

❖ ❖

Dr. Tony Bond, senior medical officer of the expedition, and his wife, the former Ivy Hazzard, of Kew, have been married for three weeks. When I spoke to Mrs. Bond she was relaxing from her duties as an air hostess in the lovely Toorak flat which she and her husband have managed to secure for a few months.

Dr. Bond, with Dr. Loewe and Mr. Law, will leave with the Wyatt Earp party some time in December.

An attractive brunette, Mrs. Bond has been an air hostess for four years; in the past 12 months on the Pacific flight. She will return to A.N.A. while her husband is away.

THE SUN, 1 NOVEMBER 1947.

4. To take cosmic ray measurements of meson intensities at sea level throughout the voyage.
5. To take oceanographical measurements of sea water surface temperatures by a thermograph attached to the sea water intake pipe in the engine room. To establish oceanographical 'stations', where possible, and to collect water samples and temperatures from various depths.

6. To take bathythermograph measurements where possible to determine temperature profiles in the ocean at various depths, using a bathythermograph trailed astern of the moving ship.
7. To collect plankton, as convenient, by hauling nets.

The programs at the two Island stations involved meteorology, zoology, botany, cosmic ray observations and, at Heard Island, geology. The main difficulty arose with the biological programs. Dakin was interested in oceanography and marine biology and we would be able to carry out a limited amount of work of this sort during the ships' voyages. But I could find no professor in the universities at that time prepared to take responsibility for the Island programs of terrestrial biology: seals, penguins, flying birds, botany.

A young biology graduate, Ron Kenny, was found for the pioneer Macquarie Island party, but no terrestrial biology was organised for 1948 at Heard Island. It was hoped that medical officer Alan Gilchrist would carry out bird observations in addition to his other duties. I adopted the role of supervisor for the biological work.

As for oceanography, it was decided that, on the *Wyatt Earp* voyage, the meteorologist, the geophysicist and I would do what we could.

The selection procedures for enlisting expeditioners were, in my opinion, quite unsatisfactory. Men were selected, as I was, on the 'old boy network' principle. Campbell would speak to an Air Force colleague or some other acquaintance and say that he needed, say, a mechanic or a radio operator. His friend would say, 'Oh, I know a good bloke—in the Navy during the War, good record, fine chap,' and Stuart would say 'Send him along'. There was little in the way of a formal interview. If Campbell liked the look of the man and he seemed keen, he was signed on.

In an interview with Campbell carried out in March 1987, ABC Radio broadcaster Tim Bowden recorded the following conversation:

Tim Bowden:	Did you personally select the men that went down?
Stuart Campbell:	Yes.
Tim Bowden:	And what criteria did you use?
Stuart Campbell:	None. As they came along, if they looked good chaps, I put them on. Some of them would turn out very good, some of them turned out good, some of them turned out very poor—you can never be sure.

Tim Bowden:	Well, I know that later directors put a lot of emphasis on screening people and, you know, making sure they were right—psychologically and that sort of thing.
Stuart Campbell:	Yes, but. . .You can't do that. I don't think you can do it. You've got to make a judgement and stick to it. Most people are all right, give them a job and they'll do it. I don't think all this. . .this psychological screening came in during the War; you can't tell what a man's going to be like. You give him a job and you let him do it.
Tim Bowden:	You must've been looking for a certain type of person though, weren't you?
Stuart Campbell:	Well, I thought anybody who had applied would be the right type anyhow. If they applied?
Tim Bowden:	Well, I suppose so. Did they know what they were letting themselves in for?
Stuart Campbell:	No, but nobody does, ever. I didn't.

My main objection to the selection procedures was the fact that no advertisements were placed in the daily press calling for applicants. I knew that there were many adventurous young men throughout Australia who would dearly love to go to the Antarctic but who, like me, did not know what jobs were offering or how to go about applying. I resolved that, if I were ever in charge of the selections, I would give everyone a chance by advertising the vacancies throughout Australia, and would also design a proper interview procedure.

Fortunately, the selection of scientists by the departments I have listed was made by their senior officers, from their own permanent officers, whose capabilities were well known.

Our storeman, George Smith, told me how he and the Heard Island cook, Norman Jones, were picked. They were in the RAAF, working at the Tottenham Stores Depot. The officer-in-charge, a Group Captain who knew Stuart Campbell, told them one morning that there were vacancies for various types of staff for an Antarctic expedition. Over a few sandwiches and a beer at lunch, George and Norman talked it over and decided to check the situation out. Since they had nothing special to do that afternoon, they went into Melbourne and along to the Victoria Barracks. Campbell looked them up and down, asked what they were doing and what they had done during the War. Then he looked at George and said, 'All right, you can start as storeman at Tottenham next Monday morning!' For Norman he nominated a much later date. Both, however, were signed on immediately.

In retrospect, I am amazed that the three expeditions of the ANARE 1948 program were successfully launched. In less than six months, starting from scratch, with no established HQ organisation, the complex logistical requirements of the Heard Island and Macquarie Island stations were provided, scientific equipment was assembled, men were recruited, outfitted and indoctrinated, and shipping and aircraft arrangements were made. Everything was extemporised using whatever existing facility could be found.

Clothing was scrounged from various ex-War supplies. It included submarine suits, camouflage windproof suits and heavy woollen pullovers from British stocks. Huts were also ex-War. There were hexagonal 'signal corps' huts from Canada, cold-weather insulated, and plywood store huts from the Australian Army, which were tropical. As soon as expeditioners arrived, they were drafted to help in the office or at the Tottenham store, where activity continued night and day and at weekends.

At port Adelaide, meanwhile, the rebuilding of *Wyatt Earp* was proceeding. In Sydney, LST 3501 was being prepared for her voyages.

It was an exciting time. Enthusiasm ran high. We were to be part of an historic venture, the first Australian expeditions to the Antarctic regions since Mawson's time. We needed no urging.

Clothing provided for those who were to sail on the *Wyatt Earp* was pretty basic. Summer in the Antarctic is not excessively cold. What is worn by skiers in the Australian winter represents roughly the level of protection required.

As well as normal Navy clothing issue, each man received a wind-proof suit, hooded parka plus trousers, made of close-woven cotton and coloured, for camouflage, with splodges of green and khaki. These camouflage suits had been manufactured in Britain for (I believe) the Norwegian campaign. They were quite effective and extremely hard-wearing. I still have mine after years of use.

Heavy, white, high-necked sweaters made from unscoured wool smelling faintly of lanoline were obtained from the British Navy, which issued them to submarine crews. They were to perform well.

Underwear consisted of standard heavyweight Australian woollen singlets and 'long john' underpants.

PREPARATIONS for the Antarctic expedition are well under way. The picture of the *Wyatt Earp* (top left) was taken in Adelaide, where she is having 16 gauge galvanised metal sheeting riveted to her planks. The upper structures of the ship are almost complete. **Top right**: Mr. P. G. Law, senior scientific officer, who will fly to Adelaide to join the *Wyatt Earp*. He was busy yesterday taking pictures of the expedition's LST at Port Melbourne. **Below left**: Mr. A. J. Lambeth, geologist, is one of the 14 scientists who is going to Heard Island. **Below right**: The LST moving from her Port Melbourne berth to North Wharf for loading yesterday. She leaves from Port Melbourne for Fremantle next week. THE *AGE*

As footwear, we had ordinary black leather Navy boots, and knee-length, wool-lined, black leather RAAF flying boots. Both were good enough for the limited use to which they would be subjected on board the *Wyatt Earp* or LST 3501, but were to prove quite inadequate for the men at the island stations. For headwear we were supplied with brushed wool Jaeger balaclavas from England, and for hand protection we used the woollen mittens manufactured for skiers in Australia. As well as ordinary Australian woollen socks, we were provided with knee-length, unscoured, white woollen socks as issued to British submarine crews.

George Smith ruled his domain at the Tottenham store with tolerant good humour, sarcasm and, when necessary, exceedingly blunt and forceful language. He was a very big man, standing six feet four inches tall and weighing over eighteen stone. No-one was ever known to offer him a physical challenge.

The ANARE Store was a large rectangular galvanised iron building standing beside a seldom used railway line, with a road on the opposite side. It had a small cubicle office at one end; the rest was just empty space.

There was no toilet. One went out the back to piss on the railway line or walked 300 yards to one of the other store huts for more serious business. One took one's lunch and drinks to work each day, arriving by electric train at the Tottenham Station and then walking half a mile to the store.

A lot of heavy handling was done there. There were no cranes or fork-lift trucks. A trailer that arrived with four 'pontoons', long plywood barges, blunt-ended, for unloading men and stores, presented a problem. Ropes were rigged linking the pontoons to the store hut. When the trailer was slowly driven away, the pontoons slid off the back and crashed to the ground.

Since George required packing cases, he would visit a nearby store hut of the RAAF complex and scrounge what he needed. One day the OC of No. 1 Stores Depot caught him at it. 'Now George,' he said, 'we can't have you doing this. Next time you need some cases give me a ring and we'll send them down to you by truck—officially!' All the authorities were sympathetic to the national objectives of the ANARE, and understood the severe deadlines that we faced. Many shortcuts were taken to avoid the tedious delays of official procedures. Helpful co-operation such as that offered in regard to the cases was the order of the day.

'Do-it-yourself' improvisation ruled at the store. As it began to fill up, organisation developed. A rat-proof, man-proof enclosure was installed at one end for clothing and the more valuable items, and a few scrounged cupboards were placed there to provide moth-proof storage. A wooden bench counter was built, across which George would outfit the men. The clothing came in only about five sizes. If it didn't fit you, George said, 'You must be deformed.'

Packing cases were filled, numbered and colour-coded—red for Heard Island, blue for Macquarie, and black for the *Wyatt Earp*. Contents were listed in triplicate on packing notes, one copy for

George, one for Head Office and one to be tacked to the outside of the case in a waterproof envelope.

Trial erections of huts were carried out and alterations effected. Marine stores were sorted and tested. Crated diesel engines, radio transmitters and cooking stoves were laboriously lugged into place, while myriads of small items—tools, engine spares, radio spares, kitchen utensils, surgical equipment and pharmaceuticals, ironmongery, scientific apparatus, and so on—arrived in a steady stream. Paperwork like invoices, receipt dockets and packing notes lists for HQ expanded. After a slow start in August, things began to roll in September. By October, the place was buzzing with activity.

It was at the store that the men became acquainted with one another. The early appointees formed a clique into which the latecomers had to break. Friendships and animosities developed but, in general, the psychological forces at work were cohesive. The men travelled, worked and drank together. They attended farewell parties and shared their friends. And, over all, team spirits began to form that ultimately would weld each island wintering party into a corporate body of enduring stability.

2

The Ship

WHAT KIND OF vessel was our ship, the *Wyatt Earp*? In 1919 a small, wooden ship had been built in Norway for herring fishing at Bolsones shipyard, Molde. She was called the *Fanefjord*. She was 136 feet long, of 29-foot beam and 15-foot draft, and 402 gross tons. She had a single deck and a fore-and-aft rig of Marconi-type sails—long, peaked, triangular sheets—and two head sails.

Purchased by the American explorer Lincoln Ellsworth in 1933, she made four summer Antarctic voyages under his leadership, in 1933–34, 1934–35, 1935–36 and 1938–39. The last was to the Vestfold Hills, where the Australian station, Davis, is now situated. Ellsworth had renamed the ship *Wyatt Earp* after the gun-toting Marshall of Dodge City, his childhood hero and a distant relative. He sheathed her with oak and steel plating to protect the hull against damage from ice.

Accompanying Ellsworth as aircraft pilot and observer was the Australian Antarctic explorer, Sir Hubert Wilkins. By the end of the voyage, Ellsworth had become increasingly exasperated by the excessive rolling of the ship. Returning to the United States from Antarctica via Australia in 1939, he presented *Wyatt Earp* to Wilkins, who promptly sold her to the Australian government for 4,400 pounds. Shortly afterwards, World War II began and the government used the ship, renamed HMAS *Wongala*, to carry stores and ammunition from Sydney to Darwin in 1939. From 1940 until 1944, she served as an Examinations Vessel in Adelaide, as a Guard Ship at Port Pirie and Whyalla, and as a Mother Ship in that region to the Naval Auxiliary Patrol. In March 1945, she was tied up in

THE *AGE*

the Torrens River, Port Adelaide, and made available to the Boy Scouts Association for Sea Scout training.

As described in Chapter 1, the Australian National Antarctic Research Expedition (ANARE) was organised by the Government in 1947 with these objectives: to set up scientific stations at Heard Island and at Macquarie Island in the Sub-Antarctic and to carry out a reconnaissance of the Antarctic coastline of George V Land to try to find a site suitable for a third station. It was decided to use a Navy LST (Landing Ship, Tanks) for the two island voyages.

Finding a ship for the 1947 Antarctic voyage had presented a real problem. There seemed to be no vessel suitable in Australia, and time was too short to search for and charter one from overseas.

Then Sir Douglas Mawson, who lived in Adelaide, remembered the *Wyatt Earp*. He suggested to the ANARE Planning Committee that she could be refitted and used. He was supported by his old colleague and captain, John King Davis, who believed that wooden ships were superior to steel ones for polar sailing.

The Royal Australian Navy agreed to be responsible for the project, and the *Wongala* was placed in the Torrens Dock at Port

Adelaide. There she was largely rebuilt, at a cost of about 150,000 pounds. Rotting timbers in her hull were replaced, and the superstructure was extended forward, to enlarge the accommodation, and upward, to provide a new bridge and chartroom, a cabin for the captain and a cosmic ray laboratory. Lieutenant-Commander William F. Cook, RAN, joined the ship in June as First Lieutenant, to attend to all matters leading to her commissioning. The ship was then just a gutted hulk on the slipway.

Raising the bridge provided her with better visibility and left more space on the main deck for cabins. An enlarged galley was fitted out, with a wardroom and a crew's mess deck, one on each side of it, beneath the bridge. The bulkheads between these areas were left unlined, so that the heat from the galley could permeate the steel to warm the messes. This proved extremely effective. A large 8-cylinder 400 HP Crossley semi-diesel engine was installed, together with two auxiliaries to provide electric power. Extra fuel tanks fitted into the hold gave the ship a cruising range of 10,000–11,000 miles at a maximum speed of $8\frac{1}{2}$ knots. An echo-sounder, a gyro-compass and a small radar set were installed on the bridge.

The sails of the Marconi rig were retained, partly as an emergency provision in the event of engine failure, but mainly to help stabilise the ship.

The First Officer, Bill Cook, was later to write:

> Perhaps we were the last of His Majesty's ships under the white ensign to use sail. I know I got a great thrill out of piping 'Hands to sailing stations' when we set all plain sail as we so often did. We found that with a favourable wind on the quarter we increased her speed by up to two knots. But the real advantage was that it slowed down her rate of roll by at least two seconds which, when she was rolling 30° to 40°, really meant something.

When the superstructure was extended forward to provide extra accommodation, what had been scuppers on the main deck were boarded over, as the hull was extended upward. This work was badly done, with gaps between the planking left inadequately caulked. These boarded-up scuppers became part of the bulkheads of the various officers' cabins. When the ship rolled enough to immerse these parts of the hull, sea water would be forced through the gaps. For much of the voyage the cabins on each side of the main deck had three inches of icy water sloshing around on their decks.

Our cabin did not suffer this disadvantage, having no exterior bulkhead. But we did not remain dry for long: the officers in the

WALKABOUT, April 1st, 1948

LABORATORY AND INSTRUMENTS

FORECASTLE AND MENS SLEEPING QUARTERS

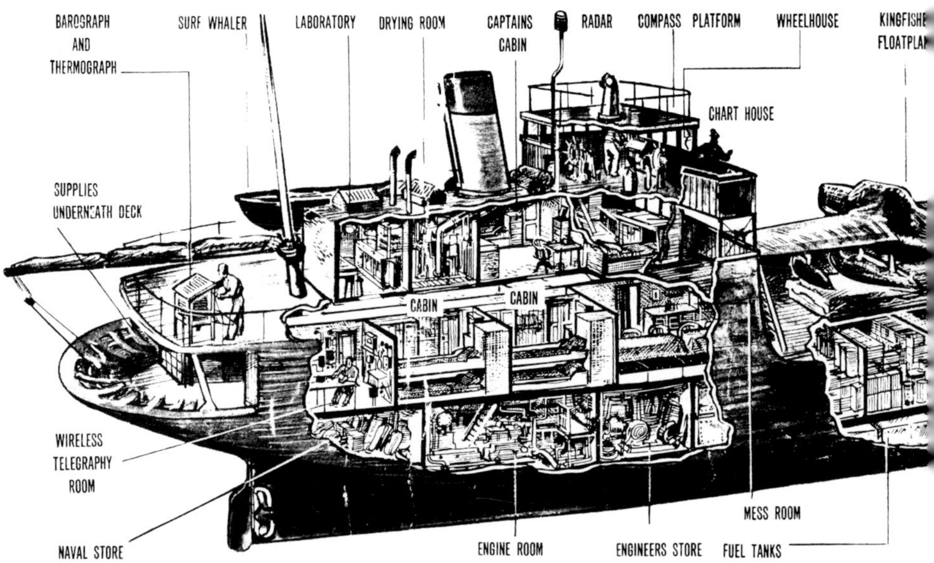

COSMIC RAY RESEARCH

In the ship's laboratory, Geiger Counters, filled with argon, are installed to register the intensity of cosmic rays, which varies with the geographical latitude. Scientists consider that this is due to the deflection of cosmic ray particles in the magnetic field of the earth. As the energy of these particles is higher than any which can be produced by man, the study of the behaviour of matter at these energies can only be achieved by cosmic ray research. Its purpose within the framework of the Australian Antarctic Expedition is to provide further information about the latitude-effect of cosmic rays as well as about their origin which is still unknown.

OBJECT OF EXPEDITION

The object of the "Wyatt Earp's" expedition is to determine one or more bases on the Antarctic continent for permanent scientific and exploratory work. Unfortunately, after her first start in December 1947, the "Wyatt Earp" had to return to Melbourne for repairs and could not leave again before February 8, 1948. Investigations have now to be confined to the eastern sector of Australia's Antarctic Territory, and it is doubtful whether the brief remaining span of the summer season will allow the party to select a suitable base. Valuable results can still be achieved, but it may be necessary to organize another trip later this year, with a full Antarctic summer ahead.

WALKABOUT, April 1st, 1948

LAYOUT OF GALLEY MENS AND OFFICERS MESS ROOMS

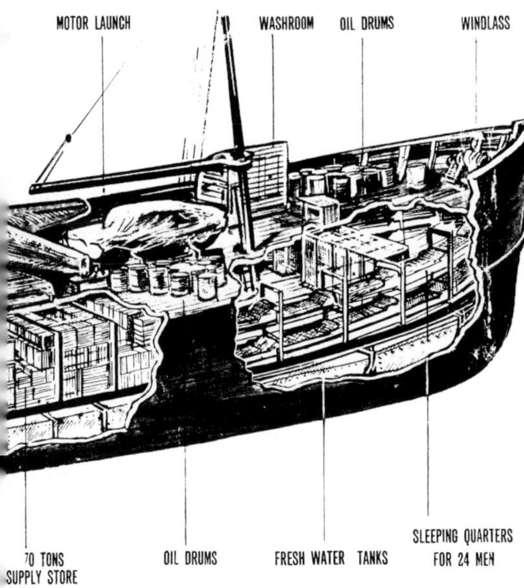

The task of the second expedition ship, "HMA. LST. 3501," is to establish sub-Antarctic bases on Heard Island and Macquarie Island. One of their main purposes is improvement of long-range weather-forecasting for Australia and her shipping lanes.

The refit of the "Wyatt Earp" for the present expedition involved the installation of a brand-new 450 h.p.-8 cylinder Crossley Diesel engine, of accommodation for 30 men, and of modern scientific equipment for measuring cosmic rays, for magnetic observations, Radar, and a Sikorski-Kingfisher plane.

HISTORY OF THE SHIP

By H. MARCUSE

On July 30, 1933, the Norwegian sealer "Fanefjord" sailed from Bergen in Norway—destination New Zealand. On board was the American explorer, Lincoln Ellsworth, and the ship carried an aeroplane in which he intended to fly across the Antarctic continent.

The sturdy little ship which Ellsworth had bought was built in 1919, constructed of wood, strong enough to weather North Atlantic gales, her bows re-inforced by metal sheathing to withstand the pressure of ice-flows and pack-ice. She was a motor vessel of 400 tons net, 135′ long, 29′ wide, with 14′ draught. With a speed of 7 or 8 knots, she had a cruising range of 9000 miles.

Ellsworth re-named his ship "Wyatt Earp" to commemorate a sheriff and frontier marshal in his home county. So began the Antarctic career of the mother-ship of the present Australian National Antarctic Expedition. The "Wyatt Earp" proved her worth on Ellsworth's four voyages to the South Polar regions, from 1933 to 1939. The distinguished Australian explorer, Sir Hubert Wilkins, accompanied Ellsworth's expeditions as organizer, meteorologist, reporter, and photographer.

The First Expedition, 1933-1934, left Dunedin, N.Z., in November and landed in the Bay of Whales in January. But the plane was wrecked and the attempt to fly across the Antarctic failed.

The Second Expedition, 1934-1935, started from Dunedin in September and reached, on October 14, Deception Island, 600 miles south of Cape Horn. To replace a part of the plane's engine, the "Wyatt Earp" made a return trip to Chile which took 18 days. But again the Trans-Antarctic flight did not succeed.

Ellsworth's Third Expedition, 1935-1936, marks one of the most memorable feats in Polar history. 17 days out of Montevideo, the "Wyatt Earp" arrived at Deception Island on November 4. On November 23 Ellsworth and his pilot started, and on December 5 they completed the Trans-Antarctic flight over a distance of 2340 miles, discovering mountain ranges of 12,000 feet altitude and other most revealing topographical features. The only mishap was a radio failure cutting off communication with the base ship. When day after day went by without news, grave anxiety was felt. The "Wyatt Earp" made a dash to Chile, returning with a relief plane, and the British exploration ship "Discovery II," was called for assistance. She reached Little America, Bay of Whales, two days earlier than the "Wyatt Earp." Here they relieved the two fliers, recognizing their tent by Ellsworth's orange signal colour (which still marks the "Wyatt Earp").

The Fourth Expedition, 1938-1939, was based on a plan to fly across the Antarctic from Enderby Land to Little America. The "Wyatt Earp" left New York in August, Capetown at the end of October. The voyage lasted 65 days, 45 of which the ship was battling through pack-ice.

Australian Antarctic Expedition. In February 1939, Ellsworth sold the "Wyatt Earp," with all her equipment, to the Australian Government for the modest price of £4400. Sir Douglas Mawson, famous Australian Antarctic explorer since the expedition 1911-14, urged the Government to employ the ship on further Antarctic work.

The War delayed future plans. In 1939 the "Wyatt Earp" was taken over by the R.A.N., re-named "Wongala." She operated in St. Vincent's Gulf, S.A., patrolling and checking on suspicious craft. In the two post-war years she served as a sea-scouts training ship in Port Adelaide.

cabin directly astern of us, which because of the fore-and-aft camber of the ship was at a slightly higher level than ours, bored holes with a brace-and-bit through the separating bulkhead at deck level in order to allow their floor water to drain through into our cabin.

The installation of a new diesel engine of much greater weight than the original small power unit was to have very serious results, which I shall describe later.

The RAN, in making the ship available for ANARE use, returned to the Ellsworth name because of its Antarctic associations. The vessel became HMAS *Wyatt Earp* on 16 July 1947.

The ship was commissioned at Port Adelaide on 17 November 1947. The commissioning ceremony set the pattern for the future fortunes of this maverick ship. In the presence of the Governor of South Australia, Sir Willoughby Norrie, Sir Douglas Mawson, the Naval Officer in Charge in Adelaide and other dignitaries, the ship was to be launched from the dry dock into the Torrens River. With the commissioning pennant flying proudly from the mainmast and the uniformed crew lined up on the deck, the chocks were removed and the little vessel slid smoothly down the slipway, stern-first into the water.

On the other side of the river was moored a coastal tramp steamer, and on the wing of the bridge lounged a crew member, cigarette dangling from his mouth and elbows resting on the railing, as he idly watched the proceedings on the opposite bank.

The *Wyatt Earp*, with momentum gathered in her run down the slipway, surged forward across the Torrens, directly towards the tramp. The cook on the *Wyatt Earp*, Stan Parfrey, later recounted the event to me:

> This fellow's idle interest gradually changed to apprehension as the *Wyatt Earp* showed little sign of slowing down. His cigarette dropped from his mouth, be began to wave his arms, then, screaming curses across the narrowing stretch of water separating the two ships, he dived for cover as the spoon-shaped stern of *Wyatt Earp* crunched into the side of the cargo ship!

One might have expected this event to have deserved some sort of naval inquiry, for the owners of the tramp steamer would no doubt have presented some sort of bill to the RAN. But as far as I have been able to determine, nothing was made of it.

The wooden hull of the *Wyatt Earp* was extremely strong. With oak beams and planking giving a total thickness of more than two feet, there were no water-tight bulkheads. In the stem of the ship the timbers were four to six feet thick. Bunks for the crew were provided in two forward sections, eight in the forecastle and sixteen in a bunkroom just aft of that. Seamen occupied the starboard bunks and petty officers those on the port side.

The thick wooden hull provided good insulation and there was no condensation on the inner bulkheads, as in later steel ships that I experienced. Seaman Norman Tame told me that, with outside temperatures around 47°F, the forecastle accommodation was about 74°F. When, in Antarctic waters, the outside temperatures dropped below freezing point, the inside temperatures were around 60°F.

The Captain's accommodation was on the upper deck, just aft of the bridge, and astern of this was a drying room, and then my laboratory.

The officers were in two-berth cabins running down each side of the main deck aft of the crew's mess and the wardroom respectively. In the centre of this accommodation block was the engine room casing. Immediately astern of this was my cabin and astern of this again, with ports onto the afterdeck, was a three berth officers' cabin accommodating 'Knocker' White and John Yule. In place of lifeboats the ship was provided with two 26-foot carvel-planked whaleboats, with oars for propulsion.

I shared my cabin in the *Wyatt Earp* with the geomagnetician of the party, Ted McCarthy. I could not have asked for a better colleague and travelling companion. A graduate in science from the University of Western Australia, Ted was about thirty years of age and had had wide field experience as a geomagnetician with the Bureau of Mineral Resources. He was used to living in rough outdoor conditions and to mixing with all sorts of men. Unfailingly cheerful and of quiet disposition, he was easy to live with and, in addition, was a tenacious and conscientious worker, and an expert in electronics.

I was thirty-five years old when the expedition started. In the 1930s and 40s, courses in physics at universities provided very little training in practical electronics. Most graduates from that period with knowledge of these techniques had begun in their boyhood with radio as a hobby, progressing from there through personal interest and self-education. This had not been my experience, and my knowledge of electronics was rudimentary. I was dependent upon Ted's expertise to cope with any but the most elementary electronic faults in our complex equipment.

Officers of the *Wyatt Earp* before departure, from left: Third Officer John Yule, Chief Engineer Frederick Irwin, Captain Karl Oom, Second Officer Norman 'Knocker' White, First Officer Bill Cook, Administrative Officer John Homewood. PHOTO BY L. LE GUAY FOR DEPARTMENT OF INFORMATION

Two other supernumeraries aboard were Dr Fritz Loewe, fifty-two years old and Head of the Department of Meteorology at Melbourne University, and Laurie Le Guay, thirty years old, the official photographer for the expedition. They at first shared a two-berth cabin on the starboard side. Le Guay became fed up with this wet cabin, and early in the voyage moved to a bunk in the sailors' accommodation, up forward. Just forward of Loewe's cabin was the wardroom, while in the centre of the ship, just forward of the engine room casing, was the galley. On the port side of this was the crew's mess. Astern of Loewe's cabin was a cabin for Irwin and Homewood, and astern of this was the radio room.

Fritz Loewe was a most interesting man. A German academic and explorer, he had come to Australia in 1939 as a Jewish refugee, and had established Australia's first Department of Meteorology at Melbourne University. He had been a member of Wegener's ill-fated Greenland Expedition in 1930–31, spending the winter in an ice cave inland on the ice plateau. He had suffered frost-bitten toes, which had been amputated.

Loewe was a familiar figure at Melbourne University, hobbling along on his toe-less feet and peering about him short-sightedly through his pebble spectacles. Tall and lean, he was singularly tough. I have never met a person so indifferent to physical discomfort. He was also a very gentle person, and an intellectual in the European sense. His memory was encyclopaedic, and his vast knowledge spanned science and history. Yet despite this he displayed an engaging naïveté and an unworldly simplicity.

Loewe was responsible, in subsequent years, for developing a small nucleus of students in glaciology, a subject not previously studied in Australia. This small cell in his Department of Meteorology was to blossom into a centre for one of the major scientific studies of the ANARE, following the establishment of Mawson Station in 1954.

Laurie Le Guay was one of Australia's best known commercial photographers, famous for his portraits of glamorous models in *Vogue* magazine. He was an inveterate traveller and a stylish travel writer. On board *Wyatt Earp* we found him to be a witty companion with a zest for life and a penchant for provocative argument on almost any issue.

Others on board were Flying Officer Robin Gray, DFC, RAAF, about thirty years of age, and an RAAF Fitter/Mechanic, Corporal R.D. ('Spike') Jones. Gray shared a cabin with the Medical Officer, Dr Tony Bond, while Spike Jones bunked up forward with the crew. Dr H. Anthony Bond had taken leave from his position as Orthopaedic Registrar at the Newcastle Hospital in order to join the expedition. He was in his early thirties. During the War he had reached the rank of Squadron Leader in the RAAF.

The ship had no facility for producing fresh water, and shortage of this commodity was a continuing source of discomfort throughout the voyage. Except while in the pack ice zone, where ice could be melted for fresh water in an improvised snow melting tank heated by the exhaust pipe from the engine, all on board were rationed to one pint of fresh water a day, for cleaning teeth, shaving and washing.

I could not understand why provision had not been made for sea-water showers and for salt-water soap. It was not until long after the voyage had ended that I found out that crew members *had* salt-water soap and could collect jugs of hot sea water each day for washing from the cooling system of the ship's engines.

There were two toilets on the main deck, one for'ard of the main mast and one aft of the officers' cabins, on the starboard side. The forward one was for the crew and that aft for the officers and supernumeraries. The forward toilet, however, became frozen up as soon as we reached Antarctic waters, and from then on the crew used the same toilet as the officers.

The ship's toilets discharged straight into the ocean from the side of the ship. The officers' toilet had a defective non-return valve and, when the ship rolled (as she did most of the time) and the outlet port dipped beneath the waves, sea water fountained up through the toilet bowl, regurgitating the contents in a disgusting fashion. It was nicknamed 'the splutterer'.

The toilets were unheated and the seats unpleasantly cold and wet from this process. An attempt by an innovative officer to remedy this by carrying on each visit a thick annular oval of felt proved ineffective, because one 'fountain' eruption of the toilet bowl would soak the felt, which would take hours to dry.

The four-berth cabin shared by McCarthy and me measured 8 feet by 7 feet 6 inches. Being in the centre of the ship, it had no portholes. Its only ventilation was an iron grating about eight inches square in the deck, just above my pillow, which opened directly on to the boat deck above. Since the deck was always wet, water dripped through on to my face until I nailed a piece of tin over the opening. Thereafter we had the choice of suffocating in our own fug overnight with the door shut, or of opening it and becoming asphyxiated by the diesel fumes which wafted in from the open engine room casing next door. The cabin was aptly named 'the black hole of Calcutta'.

The *Wyatt Earp* rolled 30 degrees each side of the vertical in an ordinary sea and, in very rough weather, rolled 55 degrees or more. She was the 'stiffest' ship I have ever known, with a period (for a double roll) of 4½ seconds. This was reassuring in one sense, for she was unlikely to capsize, but the angular accelerations generated as she moved through 100 degrees or more and back again in 4½ seconds had to be experienced to be believed. In a heavy sea, objects not securely fastened would be hurled horizontally against the bulkheads, while objects carelessly lashed would tear loose and follow the same trajectory.

Leading Cook Stan Parfrey had no relief cook. He worked from 0430 until 2130 hours for seven days a week, with an occasional Sunday off, providing meals for the whole ship's company. Various seamen were rostered, one each day, to wash up and give other assistance in the galley.

There was no refrigeration on the ship. Food on the voyage was limited to what could be carried in the form of tinned goods and dried provisions—except for meat. Surprisingly, no victuals were supplied in barrels, such as corned beef or pork in brine, or salted fish. The tinned meat rations resembled those provided to Australian troops in New Guinea during World War II—'bully beef', ham, meat and vegetables, lamb and green peas, and sausages. We had tinned butter and dried milk. The dried vegetables were of good quality, as were the tinned fruits. In general, the cook was faced with a lack of variety in the materials available to him, but he used his imagination and ingenuity to good effect, and we dined surprisingly well.

One problem he faced was that, when stores were loaded in Port Adelaide, the cases of food were stacked in the hold, along with other stores, without any system. In consequence, a number of items that appeared on his bill of lading for the provisions could not be reached until late in the voyage, when the overlying stores had been consumed.

A passage had been left between the stores in the hold so that crew members could move from their accommodation in the fore part of the ship through the engine room to the mess amidships and the bridge, thus avoiding crossing the exposed foredecks.

In an attempt to provide a source of fresh meat for the early part of the voyage, a crew's washroom on the port side of the foredeck was floored with sawdust about a foot deep, then alternate layers of ice, meat and sawdust were stacked right up to the ceiling. The sides of meat remained chilled until we reached cold Antarctic waters south of the Convergence. At that stage the meat was removed and hung from the rigging to keep cool. However, voracious skuas soon defeated this stratagem, despite Le Guay and Gray amusing themselves by practising shooting with a .22 rifle.

Oranges had been provided among the foodstuffs but they were placed in an unheated store and, before long, became frozen. Then they rotted when thawed. Potatoes also froze and became unfit to use because of unheated storage.

The RAN had earlier appointed Commander Karl Oom, RAN,

MEN OF THE WYATT EARP

Captain K. E. Oom, the commander.

AB Norman Tame, of Hurstbridge.

Mr L. Le Guay, official photographer.

Commissioned Warrant Officer J. W. Homewood, of Crib Point.

THE *AGE*

Freddie Irwin.

OBE, as Captain of the ship and leader of the *Wyatt Earp* expedition. He had accompanied Mawson's BANZARE ship, the *Discovery*, on her second voyage to Antarctica in the summer of 1930–31 as surveyor–cartographer, and had been awarded the Queen's Polar Medal. When appointed, he was the RAN Hydrographer and he brought a number of the crew of the survey ship HMAS *Warrego* with him to join the *Wyatt Earp*. At forty-three years of age he was the oldest person on board, apart from the Chief Engineer and Dr Loewe. Short and tubby in build, he was jovial by nature but behaved with the dignified reserve appropriate to his position as Captain of the ship and leader of the expedition.

The First Officer, Lieutenant-Commander William Cook, RAN, had come from HMAS *Quadrant*. He, too, was a mature officer, thirty-one years old, with a nice sense of humour and a well-balanced personality. He was very well liked by both the officers and the crew. His experience included active service in World War II.

The Chief Engineer, Lieutenant Commander Harold Frederick Irwin ('Freddie'), was an engineer from the Naval Reserve, a grizzled middle-aged man of forty-five years, with many years of seafaring behind him. Quite a competent engineer, he was somewhat

HMAS *Wyatt Earp*

PLAN OF LABORATORY

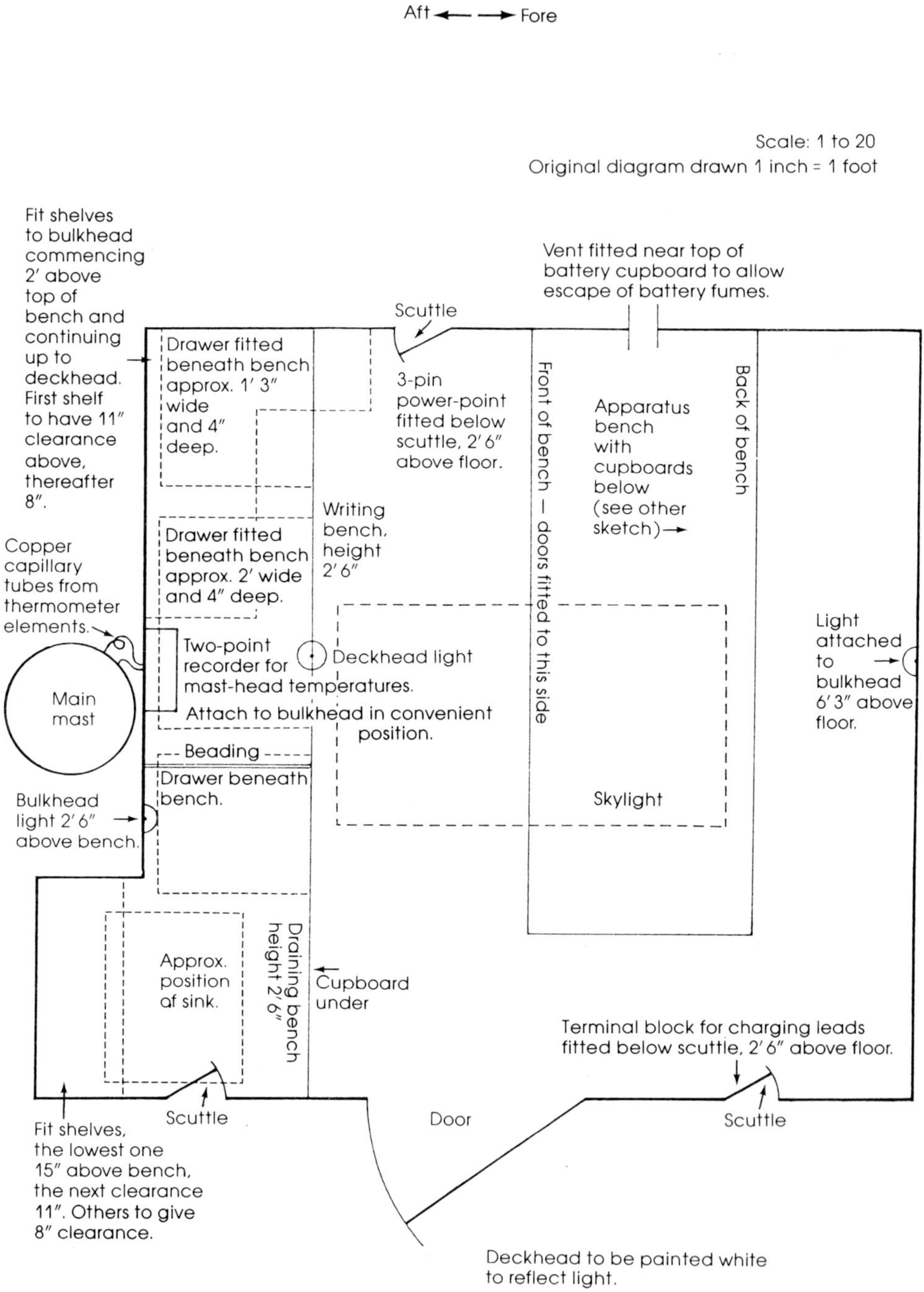

FRONT ELEVATION OF APPARATUS BENCH

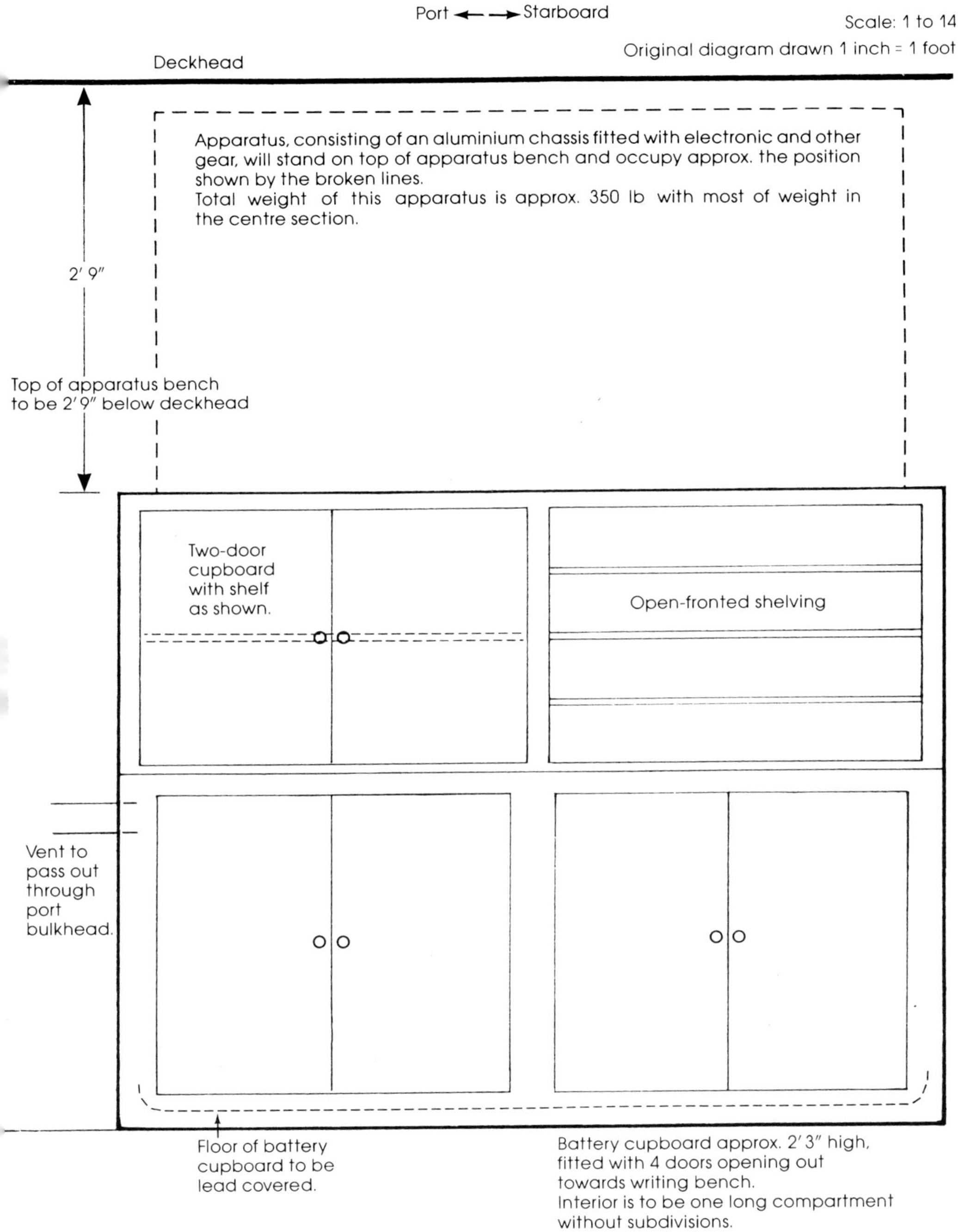

Overall dimensions of bench to be 5′ 6″ long, 2′ wide and of sufficient height to reach within 2′ 9″ of deckhead.

Designed by P.G.L.
Later built to this specification

taciturn and gave the impression of being a rather tired old campaigner. The other two deck officers were quite young: the navigator, Lieutenant Norman Harold Stephen White, ('Knocker'), had just turned twenty-five and had come from HMAS *Warramunga*, while the Third Officer, Lieutenant John Gostwyck Yule, was only twenty-one, and was from HMAS *Quadrant*. Their immaturity was in marked contrast to the developed personalities of the older, more experienced officers.

Acting Commissioned Boatswain, John William Homewood, had come up from the lower deck and, at forty-two, was a very steady, serious and conscientious officer. He arrived from HMAS *Platypus*.

The ship's crew had been carefuly selected. An annual assessment was required, for each of the three previous years, of 'Very Good' (conduct) and 'Superior' (performance). In the light of my later experience on voyages on HMAS *Labuan* to Heard Island and Macquarie Island, I would in retrospect rate the quality of the *Wyatt Earp*'s crew as very good indeed.

My job in the refit of the ship was to design a cosmic ray laboratory and to supervise its construction on the boat deck aft of the bridge and chartroom. As I was to spend most of my non-sleeping hours in this cabin, I put a lot of thought into it. The experience I had gained from designing and working in a cosmic ray hut at Mount Hotham in Victoria proved invaluable in this regard.

The plan of the laboratory is as shown. A central bench athwartships, about two feet wide, supported the cosmic ray gear which, in cabinets two feet six inches in height, occupied four-fifths of the width of the cabin. Cupboards beneath the bench held a bank of lead-acid batteries and the battery charger. On the after-side of the cabin a work-bench with a sink stretched athwartships, and a small tank that could be filled with a bucket provided water through a tap to the sink. The sink emptied into a bucket below, which had to be emptied overboard by hand. The laboratory was made light-tight so that it could serve as a dark-room for developing the films from the cosmic ray recorders.

The meson-counting cosmic ray equipment comprised arrays of geiger-counters, some shielded above by heavy blocks of lead to screen out low-energy radiation. Cosmic ray counts were recorded on mechanical counters, the numbers on which were photographed automatically once each hour on cine film. The electronic circuits involved in the cosmic ray gear were of the old-fashioned thermionic valve type, all this being before the age of transistors.

3

Cosmic Rays

IN 1946 A SMALL group of post-graduate students in the Physics Department of Melbourne University had begun a program of cosmic ray research, under the direction of Dr Henri Rathgeber, a refugee scientist from Nazi Germany. I had, in 1940, written a theoretical thesis on cosmic rays as part of my work for my MSc degree and, because of this interest, had been included in Rathgeber's group.

With the advent of the ANARE, my Department head, Professor Leslie Martin, proposed that cosmic ray experiments should be included in the work at Heard Island and Macquarie Island, and that cosmic ray equipment should also be carried on the *Wyatt Earp*. As the Lecturer next senior to Rathgeber, I was given the responsibility of ensuring that the three sets of cosmic ray equipment required would meet the deadlines set by the departures of the expedition ships.

The study of cosmic rays had been a very recent development in physics. The first indication of the presence of this complex radiation came in the first years of this century, when physicists detected ionising radiation that not only increased with altitude above sea level but which contained a constituent that could penetrate deep into the earth.

In 1929, experiments proved that much of this radiation was of a corpuscular rather than a wave nature: that is, it was composed of charged particles, such as electrons, and not x-rays or gamma radiation. At about the same time, experiments showed that the radiation was about 30 per cent greater at the Poles than at the

Equator. At sea level the intensity of the radiation is approximately constant from the Poles to about latitude 50° but, from there, it steadily declines as the Equator is approached.

In 1932, cosmic ray observations led to the discovery of the positive electron, or positron, and later work in that decade exposed the existence of mesons, a group of particles of mass between 100 and 400 times that of the electron. Although unstable and possessing extremely small life periods, these mesons had great penetrating power: they could penetrate a metre or more of lead.

Our experiments were concerned with the investigation of mesons and consisted of arrays of geiger counters designed to detect and count the numbers of mesons proceeding downward through the apparatus. Blocks of lead were placed above the counters to absorb other constituents of radiation (electrons, protons and photons) that did not have the same penetrating power as the mesons. The experiments at the island stations were to be concerned with rather complex long-term variations (over years) of the cosmic radiation, while the equipment to be carried on the *Wyatt Earp* was to measure the latitude variation of meson intensities at sea level,

From left: Joe Jelbart, Phillip Law and Fred Jacka.

particularly while traversing the 'knee' of the intensity curve at the 50°S region of latitude.

The cosmic ray team in the Physics Department comprised Dr Henri Rathgeber (research fellow and supervisor of cosmic ray studies), myself (lecturer and organiser of the Antarctic projects) and David Caro, Ken Hines, Fred Jacka, John Jelbart, John Prescott and Charles Speedy, who were post-graduate students. Jacka and Jelbart were to go to Heard Island, Hines and Speedy to Macquarie Island, and I to Antarctica on the *Wyatt Earp*.

It was decided that if a hut were to be built at Mount Hotham in north-eastern Victoria it could be used to test the equipment under frigid weather conditions. This would also give the expedition members an introduction to the sorts of climatic stresses they would face down south. Accordingly, Rathgeber and I drove to Mount Hotham on 4 April 1947 to choose a site for such a hut. There was no snow and we drove to the door of the lodge of the Alpine Club of Victoria, of which I was a member. Comfortably ensconced for that night, we wandered around the environs of the eminence named Little Higginbotham, and finally chose a spot for the hut just off the alpine highway, adjacent to the Davenport ski run. Its altitude was 5,700 feet above sea level (compare the Mount Hotham summit of 6,100 feet).

In consultation with my colleagues I had prepared a design for the hut and, on the way to Hotham, Henri and I had called at Bright and discussed its construction with Colin Heather who, with his brothers, had undertaken to build it for us. Now, on the way back to Melbourne, I authorised Heather to proceed, urging him to give high priority to the job, for winter would close in at the end of April and the road could be blocked by snow at any time after that.

We had to apply to the Victorian Forestry Department for 'permissive occupation' of the site and this I did immediately upon our return to Melbourne. We should, of course, have waited for approval before building the hut, but I considered that if we did this then the winter would beat us to it.

I purchased certain hut materials in Melbourne and sent them up by Department of Munitions transport. The Heather brothers took the rest of the materials up from Bright. They finished the building on 17 April, after four days' work under execrable conditions: snow, sleet, fog and rain.

In May, an inspector from the Forestry Department visited Hotham to inspect the site in connection with our submission for a permit. He was astonished to find the hut already built! Only the national importance of the expedition and some fast talking on my part got us out of trouble.

We managed three more car trips with hut equipment between the end of April and 4 June, although the last one had to approach Hotham from the Omeo side because the road on the Harrietville side was already blocked by snow. We were then left with the problem of taking in the cosmic ray apparatus—which was not yet completed.

The team at the Physics Department worked night and day to complete construction of the electronic meson-counting apparatus and its associated gear. Early in August, it was ready for testing at Hotham.

On Friday 8 August, Bob Dovers, a member of the Heard Island party, and I went by car to Harrietville, where we left a consignment of food for the hut with Erik Johnson, instructing him to sledge it up at his convenience. Erik was a Norwegian who, during winter, made a living packing stores and ski gear from Harrietville to Mount Hotham. We also made arrangements with Erik for the transport of the cosmic ray gear from Harrietville to the hut a week later. We then rode horses up the Bon Accord Spur to a wooden refuge hut just below the snow line, corralled our horses there, and skiied on to the Cosray Hut at Hotham to check supplies and repair the motor generator.

This accomplished, Dovers and I left the Cosray Hut at 2 p.m. on Sunday 10 August, to return to Melbourne. It was pouring rain as we fastened on our skis and, by the time we had skiied half a mile around the snow-covered road to the lodge of the Alpine Club of Victoria, we were soaked to the skin. We called in to towel off most of the surplus water in front of a roaring fire and to drink some hot soup provided by some of my fellow club members, who were enjoying a ski holiday at the Lodge.

Half an hour later we skiied along the road again for a further mile to the Mount Hotham Chalet. It nestled on a saddle at the foot of the last steep rise to the summit of the mountain. The toe iron of Bob's ski binding had loosened, so we called in at the Chalet, again half-drowned, to effect a repair and thaw out. Time was passing and we had to get off the mountain that night, so we pushed on through a heavy downpour and a strong north-west wind, on a route that left the snow-blocked road and followed snow poles up over the summit of Hotham and down the other side.

Halfway up the slope the rain changed to snow and the wind increased. Our wet clothing froze on us, and our outer windproofs crackled each time we flexed a knee or elbow. Luckily we were well clothed in wool which, although wet, still preserved some insulating qualities. Our wind-proof outer clothing also cut back the wind chill factor. Nevertheless, we began to get cold. We had fitted skins beneath our skis to aid climbing and were now plodding up the steep slope, heads down to the wind, as fast as we could move. We were unable, however, to generate enough body heat to keep warm. We each had trouble with our old-fashioned ski bindings, fumbling awkwardly to fix them with numbed fingers. Over the summit the snow had iced up and we found difficulty keeping our feet in the blizzard. We were getting colder and colder, and hypothermia had become a very real danger. Skiers have died on Mount Hotham because rain has soaked their clothing in cold conditions.

Fortunately, I knew of a refuge hut, the Diamantina, on the downward slope on the other side of the summit, so we headed for that. Luckily it was occupied, and the skiers dragged us in anxiously and stripped off our frozen clothing before a large log fire. We were blue with cold and shivering uncontrollably, so they massaged our naked bodies with rough towels and fed us hot tea and biscuits. From our rucksacks we dragged out dry underclothing and fresh pullovers and socks.

From the hut we had only two more miles of skiing along the Razorback, a sharp summit ridge flanked with a snow cornice, then half a mile down the Bon Accord Spur to the Bon Accord Hut. There we could take off our skis and mount the horses we had left there. After we were warm again, we set off. The storm continued. It was blowing really hard on the spur. We arrived at the hut just as it became dark, and tried to start a fire. Unfortunately, our matches were wet and the wood was soaked, so we saddled up the horses and pressed on.

The horses followed a narrow bridle track that wound steeply down the Spur through the forest. It was pitch dark and still raining. Bob and I sat and froze again in the saddles as the horses, with their noses held close to the ground, picked their way carefully down the steep slope, over boulders and broken branches. They negotiated slippery mud and gravelled water cascades with uncanny precision. We let the reins hang loose and left it to them. Four hours later we arrived at the tiny town of Harrietville, in the

Ovens River Valley at the foot of the Alps, and returned our horses to the owner's stable. He himself did not appear.

The Harrietville Hotel, a small and decrepit wooden building, was in darkness. We continued knocking until the licensee appeared. The pub had a reputation for being antagonistic to skiers, and our visit confirmed it. We asked for a bed but were refused. No food either—not even a cup of tea! I pointed out how wet and cold we were, gave the licensee a brief account of our trip, and begged to be allowed to come into the kitchen and brew our own tea. Reluctantly, he agreed. We paid him and made tea and toast while he disappeared back into the woodwork.

By midnight, we were outside in our car. We stripped off our outer garments, got into pyjamas (the only dry clothing left) and donned the overcoats that we had left in the boot. We then drove the 210 miles to Melbourne, arriving at six on the Monday morning.

On Monday 18 August a party of us left Melbourne with the equipment, in a passenger car and a utility truck. Arriving at Harrietville later that afternoon, we made up a number of loads suitable for packing on horses. We then retired for the night to a guest house.

Early next morning, we met Erik Johnson. In those days there were no snow ploughs in use and the road to Hotham was generally blocked by snow for the last seven or eight miles. Johnson used pack horses to carry loads up the steep Bon Accord Spur to the Bon Accord Hut which nestled among snow gums below the tree line, and on to Johnson's Hut, above the snow line. There he would transfer the loads to a horse-drawn sled and haul them over snow for the last part of the journey over the summit. This route was much shorter than that which followed the road. He had devised rough snow shoes for his horses to use above the snow line. The whole system was crude but effective.

Johnson was a remarkable character, one of the most memorable personalities in the history of skiing in Australia. Like most Norwegians of his generation, he was interested in ski jumping and, after settling in Australia in the 1920s and centring his new life around Mount Hotham, he built a practice jump that was probably the first ski jump in Victoria.

He was one of the toughest men I have ever met. He seemed quite immune to physical discomfort. He worked long hours in the snow and lived a stark, primitive life in huts on the mountains. When I first saw his hands, I was shocked. He seldom wore gloves and, working with horses and cased stores and sledges in freezing temperatures, his hands had become cracked and chapped so that they were just a bloody mess. But he did not show any signs of noticing their condition.

I remember his leading us into his primitive hut above the snow line in order to rest in shelter and have a cup of tea. He sat us down on packing cases, swept debris from the table onto the floor with a sweep of his arm, scooped tea leaves and sugar with his fingers from the bottoms of old dirty cups, then set to work lighting a fire.

For two minutes he behaved as though berserk. He grabbed a couple of wooden packing cases, and was suddenly in a frenzy of activity, jumping up and down on them and tearing them apart with his hands, scattering splinters, wood and nails, and pieces of board all over the hut. Then, briskly gathering the pieces, he pitched them into the fireplace, stuffed some crumpled paper in among them, put a match to the heap, and within a few minutes had a billycan of water boiling for tea.

Another vivid memory I have of Erik is of his kicking his draught horse in the ribs to drive it forward when the sledge it was hauling became stuck in heavy snow. But it was no ordinary kick. Standing back ten paces from the animal, he made a running jump at it, twisting sideways so that both feet struck the horse's ribs, then twisting again to land on his feet in the snow.

On his ordinary business around the mountains he travelled with one ski. Downhill he would ski on one leg, with the other stuck out behind him, and his arms waving furiously. Uphill he would shoulder the ski and walk. He preferred to carry one rather than two skis.

He had a tough load to deal with in our cases. The smallest cases we could devise for our racks of electronic gear were two feet by two feet by two feet in size, made strongly but lightly from wooden frames and 3-ply sheeting. They each weighed between 107 and 114 pounds and made a very clumsy load when strapped to a packhorse, one each side. Johnson, however, was not concerned. At 0930 hours we set off up the steep track, walking beside the horses. The party consisted of David Caro, John Darby, Ken Hines, Charles Speedy, Fred Jacka and myself, together with Johnson and twelve horses.

The trip to the Bon Accord Hut was uneventful, apart from the need constantly to readjust the loads. We arrived for lunch. Johnson rearranged the loads for transport over the next steep pinch up to Johnson's Hut, which was half-buried in the snow. Using only two horses at a time, fitted with snow shoes, we made a number of trips between the two huts. The narrow track had been cut to provide a single file path through the snow gums, in about two feet of snow. Above and to the right was the crest of the Spur. To the left, the land feel away steeply into a deep gully. Heavy rain began falling, the snow became slushy, and we were soon all very wet.

Johnson had shown considerable skill in steering the horses past the tree trunks and rocks at the sides of the path, despite the way the clumsy cases projected sideways from the packsaddles. One horse, however, slipped in the slush. Its load hit a tree and threw it off-balance. It fell heavily off the track and rolled and slithered fifty feet down the steep slope, neighing shrilly and scrabbling with its hooves at the scrub-covered terrain. The pack came apart and the two large heavy cases moved on, tumbling and sliding and bashing against the trees and bushes. They came to rest 150 feet down. Just when things were going so well!

We plunged down through the snow. We hated to think what might have happened to the electronic valves, geiger counters and other delicate equipment inside the cases. Our feet tangled in buried bushes and broken branches. We began the job of manhandling each case up to the track. With two men scrambling above, hauling on ropes tied around the case, and two on their knees pushing from below, it took a lot of time and effort. Two hours later we carried the cases around an awkward bend and reloaded them onto a fresh horse. Fortunately the fallen one had suffered no harm, and was led back to rest at the Bon Accord Hut.

We ran into further difficulty over the last half mile to Johnson's Hut. The steep slope and the deep slushy snow forced us to unload the horses and drag the cases manually up the hill. By 6 o'clock we had most of the load at the hut and retired wet and cold to the Bon Accord Hut, to spend the night.

On Wednesday 20 August we used horses to pack the lighter equipment remaining up to Johnson's Hut. It was cold and windy, with intermittent rain and snow showers. Johnson now harnessed a draught horse fitted with snow shoes to each of two rough timber sleds. Two trips of each sled were needed to take the heavier cases from the hut to the crest of the Razorback. The work was slow and frustrating, for one horse kept shedding a snow shoe. Finally, all the heavy stores were on the ridge.

The cosmic ray hut at Mount Hotham.

The Razorback is a long spur that connects Mount Hotham (6,107 feet) with Mount Feathertop (6,304 feet). The latter derives its name from the fact that its summit is a sharp ridge which, for most of its length in winter, carries a feathered snow cornice which falls away on the north-eastern side. The Razorback, at the point where Johnson's track meets it, is broader and flat-topped. From there it falls gently for a distance, then begins to rise until it joins the steep slope of Mount Hotham.

The two sleds were packed with lighter loads. They made several trips, hauled by the snow-shoed draught horses, to the beginning of the steep slope, where further weight had to be jettisoned. More relaying of sled loads was needed. The harness of one of the horses broke and we were held up for half an hour, in rain and snow showers, until it was mended. Finally, all our heavy gear reached the top of Hotham and was restacked into two sled loads. From there it was downhill past the Mount Hotham Chalet to the Cosray Hut. Caro and I left the others at the summit and skied on ahead, to prepare the hut for their arrival.

Johnson and the others, with the two sleds, arrived wet and cold about 5 o'clock. Deep snow had buried the shelter Johnson generally used as a stable for his horses, so he had no alternative but to take them back to his hut on the Bon Accord Spur. David Caro and I skied back with him past the Chalet and halfway up the summit slope to make sure things were going all right with him and the horses, then skied back in the dark to the hut, at 2000 hours. It had been a long, hard day.

On the Thursday, David and I skied to the summit to meet Johnson, who was making a number of sled trips from his hut with our lighter stores. We continued to shuttle sleds and equipment during the day and by 2000 hours everything was at the hut. We unpacked the cases and began to assemble the cosmic ray gear. We were delighted to find that damage to the sets in the boxes that had fallen was of a repairable nature.

By Saturday 23 August, the equipment was working. Mechanical counters clicked intermittently and neon bulbs flashed as incoming mesons of the cosmic rays triggered off the geiger counters. They were to continue recording in connection with a number of testing schedules for a full month.

There still remained one difficult problem. The geiger counter array was to be supplemented by an ionisation chamber designed to detect and record the total cosmic ray intensity, both mesons and other radiations. This chamber was of spherical shape, $15\frac{1}{2}$ inches in diameter, made of steel 0.6 inches thick and weighing 180 pounds. Associated with it was a lot of electrical and recording gear. When crated, the chamber presented a formidable transport problem. It was too heavy and clumsy to be carried by a pack-horse, so the Bon Accord route that we had used for meson counter equipment had to be discarded.

Mount Hotham can be approached either from Harrietville on the western side, or from Omeo on the eastern side. The Harrietville approach is short and steep, while that from Omeo is long and gradual. Both roads are blocked by snow in winter. We finally decided to transport the ionisation chamber and its accessories by motor vehicle via Omeo to the point where the road was blocked by snow, and then to use a sledge, borrowed from the Chalet and hauled by a group of men harnessed to it by ropes, to take the gear over snow to the Cosray Hut.

Accordingly, Caro, Jelbart, Speedy and Horrie Waters, the Workshop Manager of the Physics Department, University of Melbourne, drove a utility truck full of equipment to Omeo on 8 September. Next morning they drove along the ascending Hotham road until they were blocked by a snow drift at Flourbag Flat. As the distance from there to Hotham was considered excessive, an effort was made to take the car further. They spent four hours digging through the drift and then were able to drive the car, fitted with chains, to the bottom of a steep rise called Slippery Pinch.

Meanwhile, Jacka, Prescott and Rathgeber had hauled the Chalet sledge from the Cosray Hut down the snow-covered road on the Omeo side. It took them four hours to reach the party at Slippery Pinch.

The equipment, 450 pounds in all, was transferred from the truck to the sledge and all seven men struggled in soft snow to haul it up the Pinch. By nightfall they had reached only the half-way mark, so retreated by car to the Omeo Hotel for the night. Next morning they drove out and resumed manhauling the sledge. Fortunately, the weather was good, for the hauling was desperately hard. Once over the crest of Slippery Pinch the going was easier, so Rathgeber and Waters returned to Omeo and drove back to Melbourne.

As evening approached, the five remaining men abandoned the sledge on the south side of the long spur at Slatey Cutting, and skied on to rest up for the night in the Cosray Hut.

They rose at 0500 hours next morning and, after breakfasting, skied down to the sledge. The road at this point traversed the side of a hill. Drift snow had obliterated the way and produced a uniform slope across which the sledge would have had to travel if that route were to be followed. At the first attempt the heavy sledge side-slipped and no arrangement of control ropes organised by the five men could keep it on a horizontal track.

The only thing to do was to take it directly up over the crest of the hill and down the other side. Accordingly, the load was divided and relayed half at a time, at great expense of effort, to the flat section beyond the hill. Another awkward traverse was encountered at Little Higginbotham. Here the road had not been completely snowed over and enough of the flat cutting remained, through which the men could haul the sledge.

Halfway through September the tasks were completed. The exercise had proved worthwhile, for a number of faults in the equipment had been rectified, the operators had become thoroughly conversant with the apparatus, and all concerned had received a tough introduction to the discomforts and difficulties of working in snow and blizzard.

The next task was to get the stuff back to Melbourne again. After the five-man party's success in bringing the ionisation chamber equipment up the Omeo Road, it was decided to manhaul the loads in sledges down this road to cars waiting at the snow line. With this in mind, I drove up through Omeo to Slippery Pinch in a truck carrying two empty toboggan sledges. I was accompanied by John Abbotsmith, the Heard Island diesel mechanic, and Warrant Officer R.S. Swan, a RAAF photographer.

It was very windy as we left Omeo by car on the morning of Monday 15 September. As we gained altitude, the gale increased. It was soon blowing at more than 60 m.p.h. The trip was quite frightening, for branches were falling from the eucalpyts bordering the narrow road. We were concerned about being hit by a falling tree. Several times we had to stop to clear them from the track. We had to leave the car halfway up Slippery Pinch and put on skis. It then started to rain and blow even more fiercely. In the heavy wet snow we tugged and strained at the sledges and fought our way into the north-westerly gale. After two hours of exhausting effort we were little beyond the top of Slippery Pinch, with about eleven miles still to go. It was clear that we had no hope of dragging the sledges to the hut. Reluctantly, we left them at the side of the snow-covered road and skied on with all haste, to try to reach the Cosray Hut before nightfall.

It was now pouring heavily. We were soaked. The Chalet told us later the wind speed had reached 70 m.p.h. At about 1500 hours we came to a small, wooden cattleman's hut at Whisky Flat and took shelter while we brewed a billy of hot tea over a primus stove. It was a dismal place. The door swung latchless in the gale. There was no floor, only hard trodden earth which for three-quarters of its expanse was several inches under water. Half the back of the stone chimney had fallen in, and a loose flap of galvanised iron on the roof banged and rattled monotonously.

Abbotsmith wanted to stay overnight in the hut. He said we would freeze to death if we went on. I remembered my experience with Dovers and saw his point. This was not the Bon Accord hut, however, and I told him that it was more likely we would freeze if we stayed on in this horrible place than if we pressed on. We had plenty of strength left and there were only seven or eight miles to go. I advised him to push on ahead, moving as fast as he could to keep warm. He was a good skier. I would come along more slowly, husbanding Swan, who was walking for the first time on skis and making heavy weather of it.

This is what we did. We made the Cosray Hut just as it became dark. We were very wet, cold and tired. Over a good dinner, prepared by Jacka, Jelbart and Speedy against our arrival, Abbotsmith admitted he was glad he had not stayed on in the cattleman's hut.

The heavy storm had washed away a lot of the snow and left

many patches of the Omeo Road quite bare. Manhauling was therefore out of the question. Again we sought the aid of Johnson, with his big rough horse-drawn sled.

On Wednesday 17 September, I returned through Omeo to Melbourne with Abbotsmith and Swan. On the Saturday and Sunday the remaining three helped Johnson take all the gear by sled to cars waiting at Slippery Pinch. The Hotham episode was over.

The Cosray Hut at Mount Hotham paid handsome dividends for the ANARE. The operation of the equipment in this tiny hut provided the background information basic to the design of the cosmic ray laboratory on the *Wyatt Earp*, and the cosray huts at Heard and Macquarie. The tough work of transporting the equipment to and from Mount Hotham proved excellent indoctrination for the Antarctic personnel. The hut also remained there for years, to be used by ANARE and other scientists in testing a variety of scientific procedures and pieces of equipment, particularly when glaciology was developed as an arm of research at Mawson and Wilkes Stations.

On 12 December 1947, David Caro and I, with my wife Nel, drove to Adelaide in a utility truck loaded with gear. We spent five days installing one set of cosmic ray equipment on the *Wyatt Earp*, which I was to accompany first to Melbourne and then on her voyage to Antarctica. David and Nel then returned by car to Melbourne. Jacka and Jelbart took the second cosmic ray set to Heard Island, while Hines and Speedy set up the third set at Macquarie Island. All three sets functioned fairly well. A successful record of cosmic ray activity was achieved.

4

Port Adelaide to Hobart

THE *WYATT EARP* sailed from Port Adelaide for Melbourne, on 13 December 1947, defying superstition. My only sea voyage prior to this had been a holiday crossing of Bass Strait when I was eighteen years old. I remember being violently seasick and spending my one night on board curled up on a large coil of rope on the foredeck, unable to face the stench of engine-room and galley and the vomit of other sick passengers in the ill-ventilated accommodation below decks.

Things went wrong on the *Wyatt Earp* right from the start. On the 14th, the main engine failed and the gyro compass ceased to function. On the 15th, the steering chain parted. On the 16th, during life-boat practice, a davit guy-wire broke, and four sailors narrowly escaped being flattened when a heavy roll sent the 26-foot 'whaler' lifeboat crashing against the bulkhead where they had been standing. They threw themselves flat on the deck.

We butted our way to Melbourne through almost continuous storm. I was miserably seasick most of the time, but had to keep on my feet to attend to the cosmic ray equipment. Because it was newly installed, it was experiencing the usual teething troubles, so I was kept very busy. I was alone in my cabin for Ted McCarthy was to join the ship at Melbourne. At 0540 hours on 17 December, we berthed at Nelson Pier, Williamstown.

Bill Cook later wrote of this part of the voyage:

> With head winds and rough seas all the way to Melbourne it was a very wet trip. Water came in everywhere—into the mess decks and down into the gyro, which it put out of action, into the cabins and wardroom—and all this necessitated bailing.

As a trial run, the trip was very valuable. The Navy was busy for several days in Melbourne with repairs and alterations to the ship. My Physics colleagues at Melbourne University meanwhile made adjustments and improvements to the cosray equipment.

For air reconnaissance of the Antarctic coastline, the RAAF had provided a Vought-Sikorsky VS-310 Kingfisher float plane, registered A48–13. Its characteristics were:

Engine:	Pratt and Whitney R985 9-cylinder Wasp Junior, of 450 HP
Wing span:	35′1″
Length:	33′7″
Height:	14′8″
Empty weight:	3,335 lbs
Cross weight:	4,980 lbs
Fuel capacity:	120 galls
Cruising speed:	125 knots

While we were in Melbourne, this aircraft was placed on board. It was a monoplane with a large single float mounted below the fuselage. In its assembled form it stood too high to be safely transported on deck, so it was dismounted from the float, and the aircraft and the float were stowed separately in cradles on the deck.

In Melbourne, pilot Gray and mechanic Jones joined the ship's company, as also did Ted McCarthy, geophysicist and electronics expert, who was to share my cabin and assist with the cosmic ray work.

Captain Karl Oom reported that I was incurably seasick. In fact, he recommended that I should not continue on the voyage. My Antarctic career nearly came to an abrupt end there and then. I argued vehemently with him and Stuart Campbell and finally persuaded them to allow me to go on.

At 1342 hours on Friday 19 December, after entertaining the Minister for the Navy and a couple of Senators, we sailed for Hobart. Halfway down Port Phillip Bay the engine-starter handle

Left: *Wyatt Earp* setting off from Nelson Pier, Williamstown, watched by a small crowd. THE *AGE*

Below: Sailors waving to friends and relatives on the pier. THE *AGE*

Phillip Law checking over some of his gear onboard *Wyatt Earp*, 18 December 1947. THE *ARGUS*

jammed, and we stopped for two hours to repair it. Outside the Port Phillip Heads, in Bass Strait, we ran into a Force 10 easterly storm. The ship suffered a heavy bashing.

Ships designed for sailing in Antarctic waters are singularly uncomfortable to travel in, particularly on stormy voyages through the Southern Ocean. First, they are generally quite small, and get tossed around by the huge waves. Second, they have round-bottomed hulls, with no bilge keels to inhibit their performance in pack ice. Consequently they roll excessively. Third, as the ship will encounter conditions where ice forming on the superstructure will raise its centre of gravity and decrease its stability, it is generally designed to have a low centre of gravity. This makes the period of the ship's roll more rapid than that of an ordinary vessel.

Rolling through angles of 50 degrees or more each side of the vertical is not uncommon. When heavy rolling is associated with a short period of roll (that is, if the time of travel from one extreme to the other is small), the angular acceleration is great. Unfastened objects are hurled around in amazing fashion. All sorts of normal activities associated with everyday living become difficult.

One stumbles along an alleyway fending off the bulkheads with one's hands, as one is thrown from side to side. Caught unawares

Wyatt Earp in Port Phillip Bay, 1948. HERALD SUN

in a cabin by a violent roll, one learns to turn one's backside to take the impact as one is hurled across the room. Doors must be held until properly closed, for a door slamming violently when the ship rolls could take off one's arm.

Dining is particularly difficult. Wooden brackets ('fiddles') are sometimes attached to the tops of mess tables to prevent crockery and utensils from sliding around. Rubber mats are useful. Wetting tablecloths also helps to keep plates in place. But one becomes accustomed to reaching out hurriedly to grab the coffee pot or the tomato sauce bottle when there is a violent roll and, quite often, amid curses and gales of laughter, most of the articles on the tables are flung to the deck and men clutch at each other or hang on to the edge of the table to avoid following them.

A simple thing like pouring a glass of water or a cup of tea demands practice. One learns to line up the spout and handle of the teapot so that they lie fore and aft. The cup can then be moved back and forth, from port to starboard, to follow the stream of tea.

If one holds the teapot athwartships, the results can be quite disconcerting. One tips up the pot to pour—but nothing happens. Then, just as one is wondering what the matter is, tea gushes out over everything. When pouring from a glass water jug, one can see

Wyatt Earp in Hobart. PHILLIP LAW

the water heap up on the handle side of the jug, then suddenly move over to flood out over the spout.

The same problem occurs when trying to have a shower on a rolling ship. One must keep moving to follow the stream of water as it slants first one way, then the other.

Of all the ships I have travelled on, the *Wyatt Earp* was by far the worst. Its angular accelerations were beyond belief. Lincoln Ellsworth in fact became so fed up with the ship when he returned to Australia in 1939 from Queen Elizabeth Land, that he said he never wanted to see it again.

The first night out, the ship was a shambles. In my nineteen years of subsequent seafaring, I have seen nothing like it. It pitched violently, and rolled more than 50 degrees each side of the vertical, with a port-starboard-port period of four-and-a-half seconds.

Everything seemed to have come adrift. To the turmoil of the wind and the impact of the waves was added the sound of smashing crockery and the din of heavy objects sliding and bashing from bulkhead to bulkhead. Cupboard doors burst open and their contents were hurled horizontally to crash into the nearest vertical fixture. Chairs, personal gear, books, cameras, kit-bags and desk drawers bashed back and forth in the cabins. Men groaned and cursed and tried to salvage their most precious possessions.

The Captain's report stated:

> Water poured through the ship's side into the after accommodation, through the fore decking into the accommodation forward and through the bridge structure into the chart room and the Commanding Officer's cabin.

Water six inches deep sloshed around in cabins and passageways. In it floated clothes, shoes, papers, dissolving books, cardboard cartons and toilet accessories.

In the radio office a five-gallon earthenware demijohn of distilled water for batteries was trying, on each roll, to torpedo a hole in the bulkhead. The main radio receivers, which had torn loose, slid half-submerged, along with stools and chairs, on the deck. A packet of soap powder that one of the radio operators had kept in a drawer of his desk had spilled when the drawer tore loose, producing a froth four inches high on the deck water. Sloshing around in this were radio forms, log books, pens and pencils and the operator's typewriter.

The scene in the ship's galley was revolting. Pots of stew and potatoes had been hurled from the stove, despite their retaining 'fiddles', to disgorge their contents onto the tiled deck. This was covered in a cold, greasy, glutinous emulsion of sea water some three inches deep in which the cook strove vainly for foothold. Into this crashed plates, cups, cutlery and the day's supply of tinned provisions. In the crew's mess, the seats had torn loose and were sliding around under the tables, together with spilt food and plates. To cap it all, most of those on board were seasick.

The ship's library for recreational reading was housed in several cupboards fastened to the bulkhead in the passageway just outside my cabin. Doors, opening outward, were fastened by clasps, held together by padlocks. At the height of the storm, the weight of the books in the cupboards being hurled against the doors burst the clasps from the wood. The books emptied out to swill around in the water-filled passageway.

McCarthy and I lay on our bunks, hanging on grimly to avoid being thrown out. We wondered how long it would take for the storm to moderate. Lying down was easier than trying to walk anywhere, clutching for support from rail to rail, lurching and bumping from side to side along the passageway. We had tried that earlier, but had given up.

Strangely enough, we were not frightened by the chaos. In my case, this represented sheer ignorance. This was my first voyage, and I knew nothing about the sea or the vulnerability of ships. Even the experienced sailors I spoke to later, however, said they felt the same.

Lunch-time found us sitting like sand boys on the wardroom table. We were barefoot and our trousers were rolled up above the knees. We sipped soup out of warmed-up cans, while a foot of filthy water washed around under us. The whole place rocked as in a crazy dream.

And we were not even in the Antarctic! We weren't even in the Southern Ocean. We were in Bass Strait, less than a day's sailing from the Port of Melbourne!

We were eventually able to shelter in the lee of Flinders Island and, in the morning, we restored the ship to some sort of order, before proceeding. However, that afternoon, and again the following day, the gyro compass failed. At 1945 hours, on Monday 22, we reached Hobart, and berthed at the Elizabeth Street Pier.

Shipwrights immediately came on board to caulk leaks in the ship. Engineers came to repair various mechanical faults, including the gyro compass.

At 1630 hours, on Tuesday 23, the Governor of Tasmania, Sir Hugh Binnie, paid a formal visit to the ship. He was very interested in our complex cosmic ray equipment.

On Wednesday 24, I spoke to the Captain about the electric power we required to charge the batteries for our cosray apparatus. He said he could not run more than one auxiliary engine to generate power. Output capacity was already fully committed. He told me we might have to shut down our equipment if the generator became overloaded. I was taken aback, for I had repeatedly stressed our power requirements during the refitting of the ship. Apparently, they had not been taken into consideration.

I then had discussions with the Chief Engineer, the electrician and the radio operators. As a result, I proposed to the Captain that he run one generator full-time and that, on the occasions when we needed to recharge the batteries (for roughly three hours every second or third day), the ship's second auxiliary plant should be turned on. He agreed to give this a trial.

On 24 December, fresh food was delivered to the ship. Meat was stored with sawdust and ice in the crew's washplace, vegetables and eggs amidships and butter and fresh fruit in a locker space

The Kingfisher aircraft stowed onboard *Wyatt Earp*. AUSTRALIAN WAR MEMORIAL

around the rudder aft. Our departure was set for 0700 hours the next morning.

On Christmas Eve I went to a party at a Hobart friend's home and returned to the ship at 0100 hours. There was much to be done in the laboratory so I worked through until breakfast. There I learned that the ship would move out into the Derwent estuary for

Right: Christmas card sent to Dr Maurice and Jeanne Mulcahy, 1947.

Below: The Christmas dinner menu on board *Wyatt Earp*.

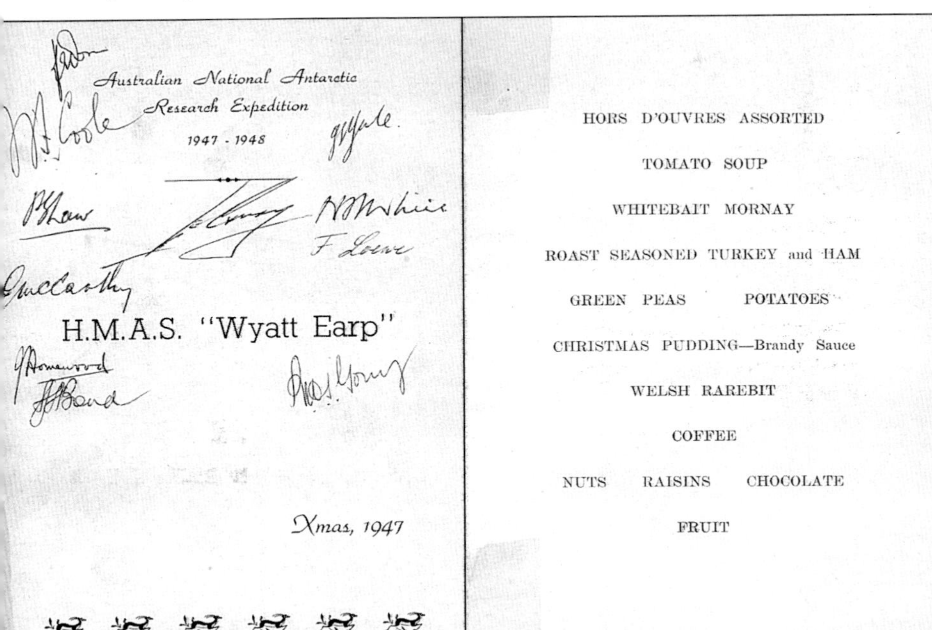

Australian National Antarctic Research Expedition

1947 - 1948

H.M.A.S. "Wyatt Earp"

Xmas, 1947

HORS D'OUVRES ASSORTED

TOMATO SOUP

WHITEBAIT MORNAY

ROAST SEASONED TURKEY and HAM

GREEN PEAS POTATOES

CHRISTMAS PUDDING—Brandy Sauce

WELSH RAREBIT

COFFEE

NUTS RAISINS CHOCOLATE

FRUIT

some hours to swing the compasses before finally departing. I returned to the lab and continued work.

Almost as soon as we left the wharf we had engine trouble—salt water emulsifying the lubricating oil. When that was fixed we adjusted the compasses and retraced our movements. The engines, however, failed to respond to the 'go astern' command at the moment of docking, and the ship continued on, to carve a swathe of woodwork from the edge of the pier with the stem, before coming to rest. Later we again attempted to depart, but the engine starting mechanism failed. Bill Cook wrote:

> So we just hung around till the engine room assessed and remedied the defect. What an anticlimax! We had done everything in our power to get the ship away from Hobart before Christmas and here we are, broken down alongside a wharf, just as Christmas dinner is announced. The gloom lasted till after the evening meal when the Captain prescribed champagne cocktails and we sang with Phil Law's accordion and danced and skylarked till about 3 a.m.

I was exhausted, and retired to my bunk quite early.

Sometime later one of the officers entered the cosray laboratory to 'borrow' a packet of Tony Bond's cigarettes, a carton of which he kept there rather than in his own cabin, because it was drier. The officer happened to open the door just as the timing apparatus operated the automatic camera that photographed the hourly totals on the geiger counters. There was a loud whirring noise, lights switched on, cameras clicked and the lights switched off. The officer got the fright of his life and fled. He believed a burglar alarm system had recorded his intrusion.

5

Heading South

THE *WYATT EARP* departed Hobart at 1310 hours on Friday 26 December 1947. Another Friday! Who said sailors are superstitious?

Bill Cook's diary records:

> In Adelaide we had several Senators and a huge crowd to farewell us. In Melbourne the Minister for the Navy and another large crowd. In Hobart a large crowd yesterday and now, finally as we sail for the Antarctic, we are not very vigorously farewelled by three men, two boys and one baby girl, none of whom seemed greatly animated by the scene.

An hour later we stopped to replace a faulty engine valve and eight hours later we ran into a Force 9 south-wester and hove-to in the lee of Eddystone Rock. We found that water continued to leak into the cabins when the ship rolled. The Captain reported:

> It was discouraging to find the sea again entering accommodation aft through the ship's side. This was believed to be due to the heavily-laden condition of the ship with its violent movement in the heavy seas producing undue strain on the superstructure.

I was at this stage feeling pretty miserable. Most of us were suffering from constipation, partly, I believe because of our subconscious reluctance to face the ordeal of the cold fountain-spurting toilet. I had, in addition, developed a painful pile. Laxative medicine I had taken only made me feel nauseated.

Fritz Loewe had the unenviable task of taking observations every four hours from a meteorological screen mounted aft on the boat

The officers' mess onboard *Wyatt Earp*. PHOTO BY L. LE GUAY FOR DEPARTMENT OF INFORMATION

deck. Getting up during the night and stepping down from his bunk into several inches of icy water were merely a prelude to his stumbling passage out onto the heaving, wind-swept deck. He clung to a stanchion while opening the screen door and focusing a torch on to the instruments.

Another task, for which he required the assistance of a second man, was to release a weather balloon once a day from the boat deck and to track it with a shipboard meteorological theodolite. McCarthy and I took turns in helping him with this. The idea was to focus on the balloon and to tilt the theodolite mounting in order to keep the horizon horizontal while one's assistant read the azimuth and bearing from the theodolite scales, at timed intervals. This was difficult and, when the ship rolled more than about 15 degrees, impossible. It was one of our most frustrating chores.

On 28 December the weather turned bad, but the wind was astern, so we made good speed. When we launched the balloon it rose ahead of the ship and spent most of its time shielded from our view by the rigging. An abortive attempt! For me the only result was a bout of seasickness.

An amusing incident occurred about this stage of the voyage. The officers' toilet in the after deck was kept locked. Loewe had used the toilet and walked away to proceed with his work. Some time later he passed the toilet and noticed the key in the lock. Thinking that, absentmindedly, he had forgotten to return it to the wardroom, he locked the door and pocketed the key. He did not know that Bond was in there! When Bond couldn't open the door, he hammered on it and shouted to attract attention. No-one, however, would release him. At lunch-time his food was passed in to him through the port-hole in the door. Later his discomfort was increased when one of the officers ignited a piece of film and pushed the smoking fragment with its acrid fumes through the same orifice. He was not let out until later in the afternoon.

Many people were having difficulty sleeping. Karl Oom's bunk lay athwartships. With the heavy rolling of the ship, he would slide down to starboard until his feet hit the bulkhead. Knees bent, he would then slide back to port until his head crunched against the other bulkhead. After a couple of nights of this he decided to sleep on the settee in his cabin, which lay fore-and-aft. When sleeping in this position one rolls from side to side. This can be reduced by placing rolled-up blankets each side of one's body.

Monday 29 December was one of the most miserable days of the whole voyage for me. To accord with naval tradition and, I suspect, to keep the 'scientists' on the ship in their place, it had been decided that all supernumeraries should take their turns with the ship's officers in keeping watch.

This was a pointless exercise, for a skilled officer and three trained seamen were already on the bridge. An untrained 'passenger' could contribute nothing worthwhile.

As scientists, Loewe, McCarthy and I had extensive scientific duties to perform. If to these were added watchkeeping tasks, then either the scientific work would not be done, or we would get no sleep.

McCarthy and I discussed this, and decided to go along with the suggested procedure on a trial basis, rather than appear to be 'difficult'. But it soon became apparent that the proposition was not only absurd, but quite impossible to carry out. Our complex apparatus kept developing faults in one section or another, and we were kept working day and night, just trying to keep it recording.

Phillip Law in the cosmic ray laboratory onboard. EDWARD McCARTHY

On the Monday I rose just before 4 a.m. to take morning watch. I found a rough sea, a wind of Force 6–7 from the SSW and the ship rolling and pitching but making little progress. Occasional rolls exceeded 40 degrees each side, with the gunwales digging under and shipping green water. I became quite exhausted trying to keep my balance and fighting against seasickness. I barely managed to last out my four hours. As I went below, at 0800 hours, I was violently ill.

Breakfast was a mess. I had one piece of buttered toast and the steward then put two boiled eggs on the table before me. Before I could grab them, however, the ship heaved and the eggs dashed madly off towards the galley, together with the sugar bowl and the jug of milk. The whole made an untidy omelette on the deck at the steward's heels. Nobody managed to collect much to eat. Those on benches between the table and the bulkhead could attempt to wedge themselves in place, but those on chairs on the other side of the table were hurled back and forth at every roll.

I crawled back to my cabin. My constipation was unrelieved, the only result of a vegetable laxative and a packet of Seidlitz Powders being a much upset stomach with indigestion and a deal of nausea. I slept for a while but couldn't face lunch, which was a game of 'chase-me-Charlie' with tins of 'M & V', the Army ration tinned 'meat and vegetables'.

An impossible situation had arisen in the galley. The 'fiddles' on the top of the stove proved to be ineffectual in rough weather, so that the containers would be thrown out onto the deck. The deck had been covered with reddish-coloured ceramic tiles, grooved to provide some kind of foothold. When greasy food had been spilt on the tiles, however, the cook and his assistant simply slid back and forth across the deck as the ship rolled. Cook Parfrey would sit on the deck holding a pot of food between his knees, and ladle out helpings into plates passed to him by his assistant.

To remedy this state of affairs, a thick manila hauser of about three inches diameter and twenty yards length was laid out on the tiles, back and forth, back and forth, to cover the whole tile area. On this foundation the workers in the galley could keep their feet. Once a week, however, the hauser would be dragged out of the galley and streamed in the sea abaft the ship for an hour or two to remove the congealed food and grease that had accumulated on its surface.

One of our geiger counters was misbehaving and McCarthy and I spent the afternoon trying to locate the electrical fault. Before we could get it going I had to break off to take the second dog-watch at 1800 hours.

After 2000 hours I went to the lab and cleaned up my pile, which had been bleeding but was getting better. I was seasick again, however, on the way to dinner. I managed to eat a meal nevertheless and, more importantly, keep it down. I had found that it is essential with seasickness to take nourishment even if you don't manage to hold on to it.

I learned that, during the day, the steering chain had parted again and the gyro had failed for the third time. Water continued to leak into the cabins as the ship rolled.

Next day my bowels cleared up and the stomach upset passed. For the first time since leaving Hobart, I felt almost normal. As a result, the motion of the ship ceased to cause me concern. Just as well, for we continued to have trouble with the counter equipment and that, plus watchkeeping, used up all our time.

At 0415 hours on 30 December, the engine revolutions were reduced because the gland where the tail shaft passed through the

Travelling through brash ice on the wing of the bridge. Bill Cook is on the far left and Fritz Loewe is on the far right. PHOTO BY L. LE GUAY FOR DEPARTMENT OF INFORMATION

hull astern was running hot. Then, at 0810, the ship stopped for repairs to a fuel pump. At mid-morning, Cook decided to try to patch the hull to prevent the worst of the leakage into the starboard cabins and radio room. The seas had moderated somewhat and he had the bosun suspend a plank over the side upon which he and shipwright Sutherland could stand while attempting to nail a large rectangle of heavy canvas over the hull just above the waterline. It was a noble attempt, for the ship was still rolling enough to dunk them both up to the waist in icy water on each roll.

After several relays of men had worked on the job they had the canvas in place. The entry of water to the cabins was decreased, but not entirely prevented. But the effort had been well worthwhile.

I was most impressed by Cook's leadership. He didn't order two ratings to do this unpleasant job while he stood on the bridge. Rather he went over the side with the shipwright and started working. By the time he had to come back on deck to thaw out, there was no shortage of men to take his place. This was typical of Cook throughout the voyage. A 'democratic' leader in the finest sense of the term, he was competent, pleasant and considerate, and was respected and liked by the crew.

That night, McCarthy and I did not hit our bunks until 0230 hours. We were both exhausted, but we had a bad night, for the wind arose from the south-east and the ship put about and went with the storm until it moderated in the morning.

Early next day I spoke to Cook and explained the watchkeeping problem. He spoke to Oom, who agreed that McCarthy and I should be relieved of such duties. Loewe, who had no major tasks except his six-hourly observations, was happy to take watches except when our team was involved in oceanographic work. He could fit his meteorological observations into the watchkeeping periods when necessary.

When McCarthy and I arrived in the wardroom for lunch, we were met with hostile silence. The Captain was not present, but the others refused to speak to us. We were being 'sent to Coventry' for presuming to stand outside the watchkeeping routine.

At dinner that night I counter-attacked. I started a spirited monologue and kept it up throughout the meal. The others fidgeted and looked at their plates. I continued addressing remarks to various colleagues in turn, with Mac's support, for the next thirty-six hours, at which stage the 'Coventry' manoeuvre collapsed. The whole episode had been bigoted and childish.

Two small kittens had somehow found their way on board when we stopped at Hobart. Whether they had been dumped there by visitors to the ship or brought on board by members of the crew I cannot say. Nevertheless they became the ship's pets and lived in the forecastle. Seaman Norman Tame adopted one and named it Chunkers. Seaman Bob Gilbert adopted the other and called it Mimi, soon replaced by the nickname 'Itsy Bitsy'. The kittens would crawl into the empty warm sleeping bags on the bunks of the sailors who were on watch and sleep for most of the day. At nights they could prowl below decks, along the alleyway leading from the forecastle past the hold to the galley and mess deck.

Chunkers developed an abscess on the jaw. Tame was most upset when one of the other sailors, without his knowledge, put it out of its misery by throwing it overboard. For the following twenty-four hours the pros and cons of this action were furiously debated throughout the ship.

On Thursday, 1 January 1948, I rose at 0730 hours to find the first really sunny day since we had left Hobart. There was a moderate westerly wind and a swell. Loewe and I flew a met. balloon but could not track it, for the ship was still rolling too much to keep the theodolite telescope view level with the horizon.

Then, at 1103 hours, the engines stopped. Upon enquiring the reason, I was informed that the new diesel engines had sunk on their beds by some fraction of an inch. The engines were coupled directly to the propeller shaft, and the subsidence was causing the shaft to bend. In an attempt to relieve the distortion, the Chief Engineer, Freddie Irwin, had loosened the gland through which the propeller shaft passed through the hull astern. When this was done, however, water leaked into the bilges and soon rose above the deck in the engine room. Part of the engine mechanism was a huge fly-wheel which had a clearance of only a couple of inches from the deck. When the water continued to rise, the rotating fly-wheel flung sea water around. There was no alternative but to retighten the gland. The propeller shaft came under increasing stress, and was in danger of snapping, while the gland overheated. Our chance of making port with only our trysails for propulsion was remote.

In the afternoon, the Captain announced that he had been ordered by the Naval Board to return to Australia. We were then in position 51°54′S, 140°31′E. So at 2030 hours we turned north, heading for Williamstown in Port Phillip Bay.

Looking back on this situation, I find it quite astonishing that it dismayed us. Every man on board wanted to continue the voyage, including myself. The Captain replied to the order with a radiogram which explained our problems in detail and tried to persuade the Naval Board to allow us to continue. A brief reply confirmed the earlier command.

Next morning the Captain again requested reconsideration of Navy's decision. This time the reply was exceedingly curt: 'Comply with my signal.' The afternoon found us making our humiliating way home. Everyone was most depressed and there were bitter remarks about headquarters' 'shinybums' making decisions in the comfort of their offices, with one eye on the politicians, without paying adequate attention to the workers facing the danger in the operational area.

The Naval Board was of course quite right! As it turned out later, the engines had *not* sunk on their oregon timber supports. This explanation had seemed unlikely to me when I first heard it, for the weight of the engine would be distributed over quite an area of the bed and, in any case, wood is fairly incompressible. It was some time after the conclusion of the voyage that I found what I think is a more plausible explanation. The *Wyatt Earp* in its early days as the *Fanefjord* in the Arctic had, for an extended period, been beset in pack ice and been lifted by ice under the hull for long enough to put a 'hog's back' bend in its wooden keel. This had remained over the years until the new and much heavier diesel engine was installed at Port Adelaide. Its weight, acting over the ensuing weeks, was enough to flatten out the bow in the keel to some extent. A few thousandths of an inch distortion in the propeller shaft would be enough!

We had spent most of New Year's Day stationary while the exchange of radiograms was proceeding. Loewe and I put the time to good use by carrying out a water sampling haul as the first 'station' of a series of oceanographic observations we had planned. Such work required a stationary ship and, in this case, involved lowering a wire carrying four water sampling 'bottles' attached at different points to collect water samples and measure water temperatures from four different depths. Winding in the wire with the hand winch was a tiring two-man task.

While the ship was drifting, the crew carried out some further canvas fitting to the hull on the opposite side. We must have just crossed the Antarctic Convergence, where the cold Antarctic water from the south meets the temperate water from the north and, at the interface, slides down under it. Over the preceding twenty-four hours there had been quite a marked drop in sea water surface temperature to 32°F as measured on the sea water intake thermograph in the engine room. On 2 January the gyro again failed.

Breakfast on Saturday 3 was hilarious as a result of heavy seas. Commander Oom had two eggs placed before him. As with me some days earlier, they crashed against the bulkhead before he could grab them. Four times the steward brought him eggs and, despite various different manoeuvres aimed at coralling the eggs, they were hurled away. The steward patiently cleaned up the mess each time.

Bond liked his eggs soft-boiled. When they arrived he played safe and placed them in his trouser pocket. On the next roll he was flung violently sideways in his seat. He pulled from his pockets a sticky mess of soft yellow yoke, cigarette packet and handkerchief. As the trousers were the only pair of casual 'civilian' trousers he had brought for lounging around in, he decided to wash them. He put them in a large cast-iron boiler on the galley stove but failed

Fritz Loewe and Phillip Law releasing a weather balloon. Watching on the bridge are, from left: Norman White, Robin Gray and Able Seaman Tame.

to watch them. The water boiled and the trousers came out shrunken, distorted and unwearable.

On Sunday 4, Loewe asked me to assist him with a met. balloon flight. We rigged the theodolite on the open boat deck aft of the cosray lab and, with the wind blowing straight from stem to stern, prepared to track the balloon as it drifted back over the ship's wake. Loewe filled the balloon from a hydrogen cylinder, tied its neck and released it. When it had gone fifty yards astern there was a report as the balloon burst. Loewe looked puzzled but, muttering something about the possibility of warm temperatures in the exhaust from the ship's funnel causing expansion of the gas in the balloon, he filled and released another. The same thing happened. I walked aft and looked below to the main deck where I saw 'Knocker' White reloading a shot gun. I said nothing, but returned to the theodolite. Loewe released a third balloon. When it, too, burst, watchers below could no longer restrain their mirth. Loewe at last realised what had happened. In his thick German accent he said; 'Vunce, it is a yoke. Tvice it is a yoke. But tree times is not a yoke. I vill not fly any more!' He then stamped off to his cabin. It was the only time I ever saw him angry.

On Monday 5, I recorded in my diary:

> Everything going smoothly. Amazing—since we received the order to return the weather has been perfect, the ship perfect and all, in general, maliciously pleasant in order to double the futility and absurdity of returning. Even the wireless room is dry!

A radiogram arrived from the Melbourne newspaper, the *Argus*:

> We advised your ship in probable path cyclone. Would appreciate from you brief message indicating your progress towards Melbourne, present weather conditions, how you expect to avoid cyclone. We have arranged for collect reply tonight.

Cook composed the reply, which read:

> Refer you James Thurber's *Secret Life of Walter Mitty*. Ta pocketa pocketa pocketa.

For the first time on the voyage we were making almost nine knots.

The following day Mac and I worked through until 0200 hours the following morning upon a part of the apparatus that had been playing up and causing many hours of work ever since we had left Melbourne. We finally got it working as the ship approached Port Phillip Heads.

Already I had come to admire Karl Oom. He was a highly competent officer and Captain. Short and squat in build, he was bland and pleasant in manner, with a keen sense of humour and a ready

Seamen battening canvas over opened seams to repair damage sustained from buffeting by Antarctic seas.
PHOTO BY L. LE GUAY

wit. Imperturbable and calmly confident in times of crisis, he inspired confidence in those he led. He would join in the fun in the wardroom in a quiet sort of way, generally retiring before the party became boisterous. I found him unfailingly courteous and considerate in his dealings with me, yet firm and definite in his decisions.

We berthed at Inner West, Station Pier, Port Melbourne, at 0930 hours on Wednesday 7 January. I scanned the pier eagerly looking for my wife Nel, and was relieved and joyful when I saw her waving from the crowed. Impatient to greet her, I did not wait for the gang-plank to be put in place, but clambered over the gunwale onto the pier and embraced her passionately. But almost immediately I had to go back on board to be introduced to the senior Naval officers who had come to greet us and to deal with several reporters who were anxious to interview me. So it was quite some time before Nel and I finally got away.

Next day, the *Wyatt Earp* moved to the Williamstown Graving Dock for inspection and refit. It was found that the propeller shaft was 47/1,000 inch out of alignment and that the wooden sheathing of the hull had been 'sprung' on both port and starboard sides by the 'working' of the ship in rough weather. In addition, there was the need to caulk the gaps in the hull above the water line that had caused the leaks into the cabins; to refit the main and auxiliary engines; to take out the propeller shaft and renew the stern bushes; to replace part of the stern post and to re-rig the steering. To remedy all these defects would take some weeks!

The next two weeks were extremely busy. I had long discussions with Trevor Heath about the progress of the programs for Heard Island and Macquarie Island while I had been away, arrangements had to be made for the return of LST 3501 from Heard Island, I had conferences with scientists going to Macquarie Island, and Mac and I had jobs to do at the Physics Department of Melbourne University in relation to our cosmic ray equipment. The time passed quickly and my enjoyment of the pleasures and comfort of my home was all too brief.

6

Uneasy Interlude

ON FRIDAY 23 January 1948, I flew to Canberra with other members of the Executive Planning Committee to attend a meeting. Judging from the attitudes of the Committee to my recommendations for scientific programs, the whole thrust on the expeditions seemed directed toward the political objectives of establishing stations and making landings. For non-scientists Stuart Campbell and Karl Oom, this was to be translated into the action of adventure and its logistical support. My experience on the *Wyatt Earp* with watchkeeping, and with the provision of electrical power for the cosray equipment, had reinforced my earlier impressions that the scientific work was being given low priority.

I therefore raised at the meeting the question of magnetic work. I asked that, if at all practicable, the geophysicist with an assistant be landed at Commonwealth Bay, or at any other approachable part of the coast, and given time to complete a set of magnetic observations. Commander Oom stated categorically that the expedition could pursue coastal reconnaissance or magnetic work, but not both. I argued that both could be managed. Sir Douglas Mawson proposed, and I supported the idea, that three men should be landed and left to carry out magnetic observations while the ship moved on to do reconnaissance of the coast. The majority objected to this on the grounds that circumstances might prevent the ship from returning to pick up this party. It was finally agreed that, if possible, three men should be landed for the magnetic work and that the ship should remain in the vicinity until the job was completed.

I could not help feeling that the prospects were still not good. In the Antarctic it would prove simple enough for Oom to use the weather or time factor as an excuse to abandon this work if he were really not keen on getting it done.

I also raised the question of watchkeeping. Stuart Campbell, who had just returned from establishing the Heard Island station, said, however, that this was a matter for the Captain to decide, and should not be discussed by the Committee.

It had been agreed that one of the expeditions should take a French observer, Yves Vallette, of the French polar exploration group Expéditions Polaires Françaises, whose activities, led by Paul-Emile Victor, had previously been in Greenland. Victor was anxious to establish a station in Antarctica and wanted one of his men to gain appropriate experience. Campbell had proposed that Vallette should accompany the *Wyatt Earp*. I opposed this because of the already crowded conditions on the ship. I was relieved when the Chairman of the Committee, Dr John Burton, ruled that Vallette should not be taken.

Campbell had not envisaged himself sailing on the *Wyatt Earp* for, according to the original ship's itineraries, LST 3501 was to have left Campbell at Heard Island when it returned to Melbourne. The *Wyatt Earp* was to have cruised westward along the whole coast of Australian Antarctic Territory and then to have sailed north to pick up Campbell from Heard Island, collect fuel from Kerguelen (which the LST had left there), and then sail to Melbourne. However, now he was back in Australia and the *Wyatt Earp* had been forced to return, he announced on 28 January that he would accompany the ship.

When I saw Commander Oom next day about this change of plan, he seemed rather annoyed. I could see that the question of leadership could arise. Oom thought that the Planning Committee should have made the decision, rather than Campbell. I felt that allowing Campbell on the ship would, from an accommodation viewpoint, be the same as allowing Vallette! Neither Oom nor I considered the cabin shared by McCarthy and me was large enough to accommodate three men on a long voyage, even though it had four bunks. It appeared, therefore, that if Campbell were to come

Wyatt Earp in dry dock at Williamstown. AUSTRALIAN WAR MEMORIAL

with us, one man would have to be put off. I was concerned lest this should be one of the three scientists.

In confirmation of my concern, Campbell asked me to inform Commander Wright (the RAN representative on the Planning Committee) that, in his opinion, there was not enough room on the ship for another passenger.

While I was speaking to Oom he said that he thought the chances of landing anybody at Commonwealth Bay (or elsewhere on the Continent) were negligible. He backed his opinion by quoting the dates on which previous Antarctic ships had left the Continent to avoid the ice. All had started in late January or early February. My later experience was to prove that mid-February was an ideal time to approach a coast rendered difficult because of sea ice. I told Oom I hoped he would leave the matter open and not prejudice the issue. There would be time enough to make a decision when we arrived in the Antarctic, for then we could see for ourselves what the conditions were like.

Shortly after this, at the ANARE office in Melbourne, Campbell took the same line as Oom. He said meteorological reports from Mawson's expeditions had shown that the weather in February at Commonwealth Bay was unsuitable for landings. In view of his earlier opinion that there was no room for another passenger, I wondered whether he was leading up to propose taking McCarthy off the ship. If this were to happen I should have no alternative but to drop out also, for it would not be possible for me to manage the cosray work unaided.

The whole *Wyatt Earp* scientific program seemed to be in jeopardy. Bond told me that Campbell had informed him that he need not do any plankton hauls—in fact, the nets and collecting jars were to be taken off the ship. Oom said he couldn't see why the cosmic ray work could not be left for a later year, when the equipment could be sent down on a large whaling factory ship. I was appalled. I told him that first, the university team had worked for twelve months to prepare the equipment; and second, the *Wyatt Earp* program was planned to complement similar work at Heard Island and Macquarie Island in the same year. Third, the following year, neither I nor any of the university team—nor perhaps any personnel from Heard or Macquarie—might be available.

The outcome of this disturbing period was that Campbell finally *did* accompany the *Wyatt Earp*. No-one was put off, and he took one of the two spare bunks in our cabin. Oom was left in absolute command of the expedition, although Campbell would clearly be able, as the leader of the ANARE, to exert considerable influence on any decisions apart from those concerning safety of the ship.

On 2 February, following completion of the dockyard work, the *Wyatt Earp* transferred to a berth at Station Pier, Williamstown. Then, on 6 February, she proceeded just outside the entrance to Port Phillip Bay, into Bass Strait, for a trial run. No better day for such an exercise could have been selected, for there was a strong south-easterly gale blowing and the sea was very rough. I was not on board, but a photo of the ship taken by a press photographer from an aeroplane showed the small vessel almost buried in a huge wave. Years later Norman Tame told me that this was the only time he ever saw a green wave come over the bows. He also described how Able Seaman Wallace was in the toilet up for'ard. As the ship was turning, she was struck by a huge wave and he was catapulted from the lavatory seat out on to the deck, bursting the door from its latch on the way. Bill Cook reported:

> The passage through 'The Rip' was very nasty. We went out with an ebb tide—a necessary precaution for a low-powered ship—and ran

> head-on into a south-east gale. The ebb against the wind produced a very short, steep sea which tossed *Wyatt Earp* like a cork. We certainly turned it on for the dockyard officers and the officers from Navy Office. The bilge pump suctions blocked and the main engine had to be stopped when the level of the water rose too high. Then, with the pumps working and the water under control, it took two hours to restart the engine.

I could just imagine the torrid time spent by all on board, as the ship rolled helplessly.

7

INTO ANTARCTIC WATERS

ON SATURDAY 7 February, with Ted McCarthy, I went on board the *Wyatt Earp* at Williamstown. Stuart Campbell joined us later. I had the top fore-and-aft bunk and the bunk under me was left vacant as a repository for personal gear that would not fit in our drawers and cupboards. Campbell took the top athwartships bunk, with McCarthy in the lower berth. The night-time fug with the door shut became even worse.

The ship departed Williamstown at 0700 hours on Sunday 8. The weather remained fine and over the next two days we logged 370 nautical miles, which was remarkably good. However, on the Thursday, with a NW wind Force 7, the ship was right back to her old form, rolling heavily and leaking again down the starboard side into Loewe's cabin, the wireless room and the wardroom.

The next night (Friday 13 again!), I was working in the laboratory. The sink was arranged to drain into a bucket underneath it, held in place by a retaining ring on the deck. When full we would take it out and empty it over the ship's side. On this particular night the bucket was three-quarters full of sink drainings, mainly developer and fixer from our routine task of developing the recording film from the cosray gear. A sudden violent roll of the ship threw the bucket out of its retaining ring and emptied its contents over the deck, flooding it with stinking slush. I mopped it up, sliding around on the wet rubber floor covering, and then had to rush outside to vomit over the rail.

A most unpleasant chore that McCarthy and I took turns in doing was attending to the batteries. These were installed in the cupboard at deck level beneath the cosmic ray equipment bench. When the batteries were on charge they gave off fumes of sulphuric acid. Attending to the batteries involved opening the cupboard doors, squatting down in front of them, unscrewing all the battery access knobs, measuring the condition of charge with a hydrometer and squirting distilled water into them where necessary. The cramped position and the stress on the stomach muscles combined with the heavy unpleasant battery fumes were generally enough to make us sick in rough weather.

Saturday was so rough, with the ship rolling 50 degrees and more, that the Captain altered course to heave-to, so we could get some sort of lunch. The gyro failed yet again.

That night there was a magnificent aurora. It was in the *northern* sky, which was surprising for we were only 55 degrees south. It followed the usual build-up to an orgasmic climax. Starting as a glow near the northern horizon that broke up into close-spaced vertical rays like faint searchlights, it developed into two parallel weaving curtains, one above the other, stretching from west to east at altitudes of about 7 and 15 degrees above the horizon. As they brightened, they displayed green, yellow and red. The depth of the curtains increased. This weaving canopy extended from over our heads to cover the whole of the northern half of the sky. Its motion became more and more turbulent until, finally, a corona formed at the zenith, writhing like a great spiral nebula, before suddenly collapsing in a few final shafts of light along the horizon.

Next morning, on visiting the lab, I found the counters rattling away like machine guns. I switched off the gear and Mac and I spent the whole morning searching for the fault. This involved unmounting the whole electronics chassis. We finally discovered that a screw holding the regulator panel in place was shorting on a wire. During this exercise, I unluckily broke off the tip of a geiger counter, which took further time to replace and solder up.

I should mention that our equipment was, by modern standards, quite primitive, although it was complex enough in construction. Since this was before the days of transistors, it was a mass of thermionic valves, condensers, resistors and connecting wires. At that time, easily replaceable panels of circuitry were unknown, and fault finding involved switching off that section, with consequent loss of record, while one laboriously probed through the wiring with multimeter leads. Further, valves would fail in quite an irregular sequence so that it was necessary to give constant attention if one were to obtain anything like a continuous record of the cosmic ray counts. Today, transistors rarely fail, and each section of the electronic system would be mounted on a small, slip-out, printed circuit panel which could be replaced while the one removed was being examined to find the fault.

The gyro was out of action twice that day.

On the Monday night there was a heavy fog and the radar picked up what appeared to be a large obstacle ahead. The ship hove-to since the navigator thought it might be an iceberg. It eventuated that it was merely a snow squall, but while hove-to the ship pitched abominably, with a sickening corkscrew motion.

Our equipment played up again the following day. Two hours of film showed no recordings, so we spent another couple of hours fault finding.

The meat was removed from the ice house and hung in the rigging. The air temperature was 35°F.

At night the cook and stewards were given the night off and wardroom personnel attended to the catering. Bond and Le Guay were the cooks, and spent the whole afternoon preparing an excellent dinner. It consisted of noodle soup, asparagus salad, pork chops with peas and potatoes, ice cream with grated chocolate and crystallized cherries, all followed by coffee, port and cigars. Irwin and Cook dressed up to serve the dinner, the Chief Engineer resplendent in pantaloons, white steward's jacket, false teeth of orange peel, night-cap headwear and medals made from can tops. Campbell, Loewe, McCarthy and I washed up afterwards. There was a huge assortment of crockery, cutlery and dirty cooking utensils. A singsong session followed, accompanied by me on the piano accordion, and the party became wilder, finishing up with an all-in water throwing battle. Bond cleared up the resultant mess.

Late in the evening, when he had drunk enough to become loquacious, he voiced to me the resentment in the wardroom of the fact that McCarthy and I had been exempted from keeping watch. This merely confirmed what had been quite obvious to us both—but I was a little surprised that it persisted.

As already mentioned, the wardroom was warmed by heat from the galley transmitted through the uninsulated intervening wall.

Officer of the watch.

On a party night, however, with eight or ten men sitting around the tables in the rather confined space for a couple of hours, the temperature would rise to more than 80°F. There were two portholes in the fore bulkhead opening on to the foredeck. A man would unscrew the fastening of one thick glass hinged 'window' and call: 'Flush'. He would then open the port wide, and count 'One, two, three, four, five'. A gale of wind—and often snow—would sweep through the room before he slammed it shut and screwed the fastener down. This performance, carried out every fifteen minutes or so, reduced the temperature to an acceptable level.

On Wednesday 18 February we sighted our first icebergs in position 65°03C, 143°35E. The transition in weather was quite sudden. Overnight the north-west wind dropped, to be replaced by a steadily increasing south-easter. The air temperature fell to 22°F, and the sea temperature to 30°F. A general air of excitement pervaded the ship, with the crew dashing about happily with cameras. We passed a number of icebergs, mostly flat topped, although some smaller ones were pinnacled. By lunch-time bergy bits up to the size of a large room were floating past. With the strong wind the chill factor was high, and those handling cameras had a salutary introduction to Antarctic conditions. When I left the lab at 0100 hours, the air temperature was down to 20°F.

Stuart Campbell's diary for that day says:

> I finished my watch (2400–0400 hours) and turned in about 0430, only to be awakened again about 0600 by a message from the Captain to say there were icebergs in sight. I didn't get up. I knew they'd still be around later on and, having just got into my bunk, I didn't relish the thought of getting out again just to look at an iceberg, of which we were sure to see plenty more.
>
> When I did get out, about 0900, I found everyone on deck getting their first thrill out of the sight of an Antarctic iceberg. And though I've seen lots before, I must confess that I, too, got a thrill out of that first sign of the Antarctic. They were all around us, at least six of them. Huge flat slabs of gleaming white ice with vertical sides rising 150 feet sheer out of the sea. Along the waterline are huge caverns in which the sea boils and foams as the surge runs in and out. Here and there the sides are split in deep V-shaped crevasses, out of which shines a pale iridescent glow of the deepest blue. Two of the bergs were at least five miles long and the others between one half and two miles.

After breakfast next morning I went onto the bridge, for all were complaining how cold it was. There was a Force 8 SE storm that built up to Force 9 later. I was somewhat disappointed to find it wasn't very cold—not anywhere near as cold as I had often experienced at Mount Hotham. Most of the others had obviously never suffered anything more severe than a Melbourne winter! Their tolerance would change considerably before we reached home again.

Campbell wrote:

> The ship is rolling heavily and I struggled into my clothes with difficulty, falling about the half-lit cabin like a drunken man, and finally emerged more or less waterproofed and windproofed from head to foot. I know it's going to be cold and it is. The air temperature is 20°F and the sea is 29°F. The deck is covered with ice as I stagger and slide along to the bridge and take over from Robin Gray who tells me that they sighted the edge of the pack ice about 0300, but it is too rough to push into it, so we are turning NW for a while and are now running along but out of sight of it in open water with a few small bergy bits and growlers floating around.

The day was disastrous for the cosray program. I went to the lab at 1000 hours and was horrified to find, on opening the door, that the cosmic ray equipment in its enveloping aluminium chassis was bashing from side to side of the cabin along the top of its bench, as the ship rolled. It was making horrible, horrible crashing noises. The retaining wooden 'fiddles' had been torn loose by the violent rolling of the ship.

At first I didn't know what to do. I looked around for some object to wedge into the eighteen inch space between the equipment and the bulkhead. Since there was nothing suitable, I used my forearm, hoping that the 'throw' of the weighty equipment would not break it. I stood there with my hand on the aluminium end-panel of the chassis, and my elbow a couple of inches off the bulkhead, and took the force of the sliding bulky object on my forearm on each roll. This stopped the equipment from bashing itself to pieces but I was stuck there, not daring to leave it to itself. I called for help—but no-one could hear me. After nearly half an hour of this, when I was wondering how much longer I could hold out, I attracted the attention of a passing sailor, who raced off to call McCarthy. He came up to the lab, then dashed off again to find a baulk of timber to wedge against the sliding equipment. We then surveyed the damage with sinking hearts. Was this to be the end of our experiment?

All the connecting power leads had been sheared off. Worse, the heavy lead slabs placed above the geiger counters to absorb unwanted soft radiation in the cosmic rays had become displaced and fallen onto the counters, smashing these delicate tubes and distributing broken glass all over the deck. Two power transformers had torn loose and smashed up some of the valves and circuits.

We looked at each other despairingly. Then we started work. At 1630 hours I was exhausted and had to give up, but Mac, who had slept in until called by the sailor, soldiered on. I slept until 1830 hours, then relieved Mac who was by then also exhausted. By 2200, between the pair of us, we had repaired the most obvious damage, but the ship was still pitching and tossing violently, and I went anxiously to bed.

Campbell's diary for 20 February has this interesting note about watchkeeping:

> 'All night in' last night. Only those who have been keeping watch can appreciate the pleasant feeling of coming off watch at 0800 hours knowing you have no more duties until 0800 next morning. No struggling to get a few hours sleep in before midnight; no getting up at 0345. Just a whole night in bed which comes around once in four days. I enjoyed 'All night in' so much that I hardly bothered to sleep, just lay there thinking how nice it was. In the early hours of the

> morning, we turned beam-on to the sea and the rolling was terrific. My bunk was athwartships and I was sliding up and down, my feet on the bulkhead one minute, my head against the opposite bulkhead the next. Even if you push yourself up one end, you still slide up and down inside your skin. But it didn't matter—I had the whole night in.

Next morning I found the ship cruising back and forth, from west to east, waiting for the storm to moderate, just outside the edge marking the northern limit of a zone of pack ice. Late in the morning the wind moderated and the ship closed on the ice edge, where the water was comparatively calm. All day and that night we worked in the lab, replacing geiger counters, screwing things back into place and soldering leads and joins. By 2000 hours the main counters were again operating.

After lunch the ship had found a break in the ice edge and had sailed roughly SSW, in a lead a couple of miles wide, through dispersed pack and brash ice. The pack grew thicker and the lead degenerated into a narrow canal a few hundred yards wide.

We glided on, with ice floes on each side, upon some of which waddled Adélie penguins. As the ship passed, and we shouted at them, they would panic, throw themselves onto their breasts and toboggan around on the floes. Pretty Cape pigeons flew about the ship and, in a pool between the floes to port, four killer whales surfaced and 'blew' before submerging again. A magnificent small glacier berg slid past close to us on the starboard side, its top encrusted with hexagonal ice blocks. Cameras clicked madly on the deck and the bridge.

The following transcript is from a wire-recording made by me on Friday 20 February as the *Wyatt Earp* entered the pack ice:

> Last night the wind blew a strong gale, force 10, and the ship pitched and rolled abominably. This morning the storm moderated and we moved into the calm water in the lee of the pack ice.
>
> Let me explain the appearance of the pack. As one sails in toward the continental land mass of Antarctica, one meets first of all large icebergs which, because of their size, are able to drift a long way north before melting. Further south the sea has, bobbing about in it, small irregular pieces of ice, remains of ice floes or bits broken from the bergs.
>
> Further south still one meets ice floes, flat chunks of ice varying in size from a few square yards to half an acre or so across the top. One or two feet of their thickness is above water and the other five or ten feet are below. The tops are perfectly flat.

At first they are wide apart, with large areas of open water between, but as one sails nearer the land, the open spaces become smaller and degenerate into twisting, narrow spaces between the drifting floes. Because of the enclosed nature of the water between the floes, the sea is as calm as a lake. Ultimately the sea may become completely covered with tightly packed ice floes. In such stuff a ship may be trapped.

So far we have made two attempts to reach Commonwealth Bay through the pack ice but, each time, about thirty miles from the coast, the floes have closed in so much that a clear way forward could not be seen and we had to withdraw.

As a rule there is open water, free of pack ice, immediately adjacent to the coast, but from the crow's nest, the limit of the ice on the shore side could not be seen.

It is extremely pleasant cruising through the pack. The ice fragments, penguins on the floes to amuse us and killer whales, sporting like dolphins in the water and blowing high into the air, add further interest.

The penguins are mostly Adélies, standing about two feet six inches high. There were more than a dozen on one floe, waddling about in their quaint, upright human manner and gazing curiously at us. We sailed right up so that the ship grazed the floe, with our official photographer, Laurie Le Guay, at the bows ready for a shot.

When we drew alongside, we shouted to make them move. First they gathered together in a bunch and milled around in a panicky sort of way. Then they headed for the water, some waddling and some tobogganing. This latter motion is quite amusing as they throw themselves forward on their breasts and slide, with their legs pushing on the ice behind to keep them going and their flippers paddling in front to guide them.

When they got to the water's edge we hoped they would go in, but no—scrambling to their feet again they looked over the edge of the floe, then milled around in a cluster, then looked at us, then looked at the water again, for all the world like a group of bathers squibbing it on a cold morning. Some even tried to push a comrade in! By this time we had drifted away a little, and they decided to stay where they were.

On 21 February Campbell wrote:

A very comfortable night last night with the alleyway doors open and plenty of fresh air in the cabin. This ship has the unusual characteristic of taking water over the stern in a rough sea, which means that for safety the alleyway doors have to be kept closed at night, otherwise there's a good chance of a wave coming over, running down the alleyway and flooding all the cabins and wardroom with a foot of water. If you keep the doors shut, however, there is hardly any ventilation and the accommodation space becomes full of fumes

> from the diesel engine and you wake up with a headache and a hangover in the morning. This soon goes off after one has been on deck for an hour but, whilst in the open sea, there are only those two alternatives—a hangover in the morning, or a good chance of a flooded cabin in the middle of the night. We usually choose the hangover.

For the next couple of days we weaved around on the fringe of the pack ice zone, in and out of leads, with frequent snow storms, without finding any way through. It became clear that the Captain had no intention of pushing into the pack itself and that, unless an open lead could be found to the coast, we would not be able to make a landfall. We were at about 66°30S 146°00E.

As a consequence of the rough seas the ship had negotiated, spray cast up by the bows had frozen and coated the forecastle, foredeck and rigging with a thick layer of ice. The recent heavy snowfalls had built upon this, with the result that there was danger of the ship becoming unstable. Ships have been lost in the Arctic when they have capsized due to a 'top hamper' of ice built up in cold conditions by spray in stormy weather. So the Captain ordered all hands to work to clear the ice away.

Officers, sailors and supernumeraries worked with picks, axes, spades and shovels to chip the ice off the deck and heave it overboard. It was heavy work and, although the air temperature was 20°F, and we were stripped to trousers and shirts, we were still sweating. At the end of two hours, the job was done. The sailors then hosed the debris off the deck with hot salt water from the engine cooling system.

Mac and Loewe and I noticed a considerable change in attitude of the others on board towards us at the end of this task. We were fit and strong and eager for some exercise and had thrown ourselves vigorously into this job, more than matching the efforts of the seamen. This apparently surprised the non-scientists. Had they assumed that we were incapable of such work, or did they think scientists would consider it beneath their dignity to stoop to such labour? Whatever they thought, we were accepted thereafter as almost normal members of the ship's company.

The Captain tried charging one of the smaller floes to test the ship. I had thought we might crack it and go through, but the ship had neither the weight nor the power: it merely bounced off. As an ice breaker, the ship was a gnat!

On Saturday 21 February we all worked helping to attach the large float to the Kingfisher aircraft and prepare it for flight. This took Gray and Jones about five hours of hard work. The aircraft was then secured with about six heavy tackles to prevent any movement until there was an opportunity to fly it.

Campbell noted:

> By dusk the main assembly was complete and she is now sitting up on the hatch ready to go, except for a few finishing touches. I confess I made a mistake in selecting this type of aircraft for this ship, as she is far too big for us to handle comfortably. What is needed is a small aircraft that can be kept ready for flight at short notice.

To the south, beyond the pack, lay the Mertz Glacier Tongue. There were many other icebergs that had obviously broken from it. There must have been an emperor penguin rookery to the south somewhere, for we saw many of these beautiful birds standing motionless on the floes.

As the *Wyatt Earp* could not generate fresh water, we had been severely rationed. Apart from drinking water, each man was allowed to draw a jug of fresh water once a day for his personal hygiene. Most of this went in washing one's face and hands and cleaning one's teeth. If one went sparingly and accumulated a small store in a bucket, one was able to have a kind of wipe-over 'sponge bath' every few days, but there was no possibility of washing any clothing. I was looking forward to a chance to land on an ice floe and chip off a ton or so of ice to place in the ice melting tank that had been fitted to the ship.

On 22 February, Campbell wrote:

> There appears to be little point in hanging around this area at present, so we agreed to go east and try to get into Fisher Bay, or to the base of the Ninnis Glacier or, if that were not possible, to carry on east to 160° longitude, the limit of the Australian sector.
>
> As the coast has never been seen, except from the air, and has not been approached so late in the season before, we might be able to do some original survey. After reaching 160° longitude we will go on to the Balleny Islands to survey them and collect geological specimens and then return to the Commonwealth Bay area and have another attempt to get in.
>
> Today our southerly course was blocked by heavy pack, but we cannot say access to the land was impossible. We didn't try enough. Ten miles east or west we might easily have struck open water, but personally I think this very unlikely.

Stuart Campbell would adopt various forms of behaviour to demonstrate how physically tough he was. Long before the days of jogging and the present concern with fitness, he would talk of his

Wyatt Earp moored to an ice floe for 'watering ship'. PHOTO BY L. LE GUAY

early morning cold swims in winter, and his long-distance runs. On the ship, he would boast that he never ate breakfast. At morning tea he would put away a meal of sardines and biscuits. On the voyage he wore the same clothing (singlet, underpants, shirt and all), without changing. For weeks he would spurn washing his body. Perhaps he was just setting an example in saving water—but after a few weeks our cabin positively stank. This was another reason I looked forward impatiently to the day when we would be able to 'water ship' from a handy ice floe.

In this context I was interested recently to read the following note in *Mawson's Antarctic Diaries*.[1] During the voyage of the *Discovery*

1. *Mawson's Antarctic Diaries*, edited by Jacka and Jacka, Allen & Unwin, 1988.

'Watering ship' at an ice floe are, from left: Bill Cook, John Homewood and Freddie Irwin. PHOTO BY L. LE GUAY FOR DEPARTMENT OF INFORMATION

on the BANZARE Expedition of 1929–31, Mawson makes the following entry:

> We [Mawson and his Captain, John King Davis] had a long talk about the scientific staff and water. He rightly contends that Campbell is dirtier than he need be. I pointed out that there is no excuse for such dirt but stress the fact that Campbell is a hard worker and on dirty jobs, as in the wells of the Echo Sounder.

On Monday 23 we alleviated the water problem a little by tying up to an ice floe and filling buckets with ice. All enjoyed the chance to stretch their legs off the ship. Robin Gray tried to get the aircraft going but the engine died each time he started it—too cold, apparently. When we had finished collecting ice, some of the ratings played around on the floe with a medicine ball. Two who crossed to another floe to look at a seal basking there were stranded when it drifted away. Then everyone was called back on board, and the ship cast off and sailed after them. Eventually they were able to climb back on board.

Proceeding north-east we ran for several hours along a line of huge grounded icebergs. Deep caverns in their vertical cliffs glowed with the usual shimmering blue light and the sea swell, rolling into them, compressed the air within, which then erupted with a roar of sound, and cascades of spray.

Tuesday 24 was a memorable day, dawning bright and sunny after a frosty night when temperatures dropped to 12.5°F. This was the first real period of sunshine for more than a week. In the afternoon we sailed through the icebergs lying offshore from the Ninnis Glacier Tongue. The scenery was indescribably beautiful. The blue sea was calm, rippled by a gentle breeze, and the ice floes glistening dazzlingly in the slanting sunshine. Around us, as far as the eye could see, were huge 'domed' icebergs, apparently about 66°28S, 148°53E. For a short time we then pushed south but thick pack ice at around 67°10S forced us to retreat north. A flight here would have been valuable—but, yet again, the engine was too cold to start. Campbell noted: 'The heating device is quite inadequate and the portable Stewart Warner Heater that I had taken so much trouble to get had been offloaded in Adelaide as "not considered necessary".'

I had a long discussion with Fritz Loewe about various theories of the formation of domed icebergs. This problem has continued to interest me throughout my Antarctic career. Observations of these bergs at different places along the coast of Antarctica have provided me with information upon which to base a considered opinion.

The common 'tabular' iceberg, as its name implies, has vertical sides and a flat top. A 'domed' iceberg has a top that rises gently from every side of the berg. The summit of such a dome can be many times higher than the ice-cliff edges of the iceberg. Frequently the domed surface is riven with large crevasses, quite visible from a nearby ship. The sides of a 'domed' berg slope outward, the tops of the ice cliffs reaching out over those parts that are at sea level. Generally, soundings become shallower as a ship approaches a domed iceberg. It appears that such a berg is grounded.

The obvious theory is that such icebergs are not only aground but have been for many years. Over these years snow has built up on top to produce the heaped 'mound', while glacier-type ice flow from the centre of the top surface out toward the edges deforms the vertical edges to produce the outward-leaning appearance. The icebergs off the terminations of the Ninnis Glacier and the Mertz Glacier are fine examples of the domed type.

The subject is, however, a little more complicated than this. The domed berg of this type is symmetrical: that is, it appears domed from whichever side one views it. My theory was to be complicated in later years when, in the Dalton Iceberg Tongue and the Dibble Iceberg Tongue, I was to observe icebergs that were domed when viewed from the front or back, but flat ridged when viewed from either side. Although these were also grounded, aerial photos of the glaciers from which they had broken off suggested that their shape had not developed from snow accumulation and subsequent outward ice flow.

For some miles inland from the point at which these icebergs had broken away from the Antarctic ice sheet, the sheet had become 'humped' quite visibly, as it passed over continental hillocks or islands close offshore. Thus, from the air, one could see, inland, the smooth descending continental ice sheet and then, close to the shore, numerous humps and undulations, with heavy crevassing. Finally, further out, beyond the coast, these humped parts of the ice sheet were detaching themselves by breaking off along the lines of least thickness, producing 'domed' bergs. But these had *vertical* edges and they were domed in cylindrical, not spherical, fashion.

I had provided on the bridge two log books, one for recording birds sighted and one for whales. Accompanying these were a bird book, to assist in identifying the species, and a chart showing the characteristics of the various kinds of whales. Watchkeepers were encouraged to take an interest in making observations and recording what they saw. One entry (made, I suspect, by Freddie Irwin) read as follows:

> One blue-beaked beetle-browed blowsy-bosomed brown-bottomed black-backed bright-breasted big-bellied baby bludger bird.

Still unable to push toward the coast, we turned north to leave the ice from our position of approximately 66°35′S, 148°07E. On the way we ran into trouble: loose pack ice had drifted across between the domed icebergs and blocked our path. For the first time the ship had to 'work' the ice, and she shook and bumped as she pushed the floes aside to get through.

The sun set beyond the domed bergs in a glory of colour, tinting them and the pack with delicate hues of pink and gold. Then, at about 2100 hours, the moon rose, and the scene became even more beautiful. The low full moon, huge and orange-red, shone slantingly across the ice floes, turning to diamonds the ice crystals forming on the surface of the water between them. The ship slid smoothly along through a rose-coloured vista of unnatural shapes of vast magnitude. We leaned over the rails and watched enthralled. By midnight we had left it all behind, as we headed north through open water.

Next morning we turned east towards the longitude of Cape Freshfield. In contrast to the day before, the weather was dull, grey and depressing, with occasional snow showers. The gyro went 'off the board' once more.

On Thursday 26 we passed numbers of humpback whales, and soon afterwards sighted a whaling fleet. The mother vessel was the British factory ship SS *Balaena*; she was accompanied by a refrigerator ship SS *Ketos*, and two chasers. There were obviously other chasers elsewhere.

We sailed through large quantities of freshly formed 'pancake ice' blown out by the SE wind in thin lanes across our path.

At night we could easily have imagined we were back in civilisation again. The sea was oily calm and there were five rows of lights around the horizon, fixing the positions of the whaling ships.

Robin Gray had at last managed to start the aircraft engine, after keeping the heater going continuously for more than twenty-four hours.

On Friday 27 we headed south again and reached almost 66°45′S —our best yet—at longitude 157°22E. We were blocked at 1400 hours by heavy pack ice extending from 100 to 290 degrees. The ship turned and headed north out of the ice again, and the Captain announced that we would proceed to the Balleny Islands, about 150 miles north-east from where we were. Mac and I hoped we might be lucky enough to get ashore to make geomagnetic observations.

8

THE BALLENY ISLANDS

AS 1500 HOURS on Saturday 28 February, we sighted the Ballenys when the mist through which we had been sailing lifted. They were forbidding in the extreme. They consist of four main islands —Young, Buckle, Borradaile and Sturge—and two very small ones, Rowe and Sabrina. Young, the largest, is over 20 miles long. All are of volcanic origin and permanently ice-covered, with precipitous coastal bluffs up to 1,000 feet high and glaciers which terminate in ice cliffs 50 to 100 feet high. We cruised around looking for likely landing spots, without success.

After dinner the ship stopped, and Loewe, Campbell, McCarthy and I did a hydrological station, collecting six water bottles from different depths. We finished at 2330 hours. Mac and I then worked on cosray films until 0100. During the hydrological exercise we saw a number of humpback and blue whales.

The Balleny Islands were discovered on 9 February 1839 by Captain John Balleny of the ship *Eliza Scott*. This was one of two ships despatched from London by the British firm of Enderby Brothers to investigate new whaling and sealing grounds in the Antarctic. Captain Freeman was master of the second ship, the *Sabrina*. He rowed ashore in a long boat and landed at an unspecified point on Borradaile Island, but he stayed only for a few minutes for his boat was being battered by the surf. The description of the

Rowing ashore, Balleny Islands, 1948. The three rearmost oarsmen are, clockwise from back left: Able Seaman W. R. Wallace, Phillip Law and Able Seaman Tame. PHOTO BY L. LE GUAY FOR DEPARTMENT OF INFORMATION

spot tallies with the place where we landed for an equally brief period.

Balleny's expedition made rough sketches of the islands and named the major ones after the Enderby partners. Approximate positions were delineated and major features named. Like us, they were unable to survey the southernmost island, Sturge, which was surrounded by heavy pack ice. They were the only people who had landed before we had.

The first accurate chart of the Ballenys was made by RRS *Discovery* which, as part of a series of Discovery Expeditions carrying out oceanographical work in the Southern Ocean, visited the islands in February 1936, and again in January 1938. No landings were made. On board the *Discovery* in 1938 was an Australian meteorologist, Alan Cornish, who was a contemporary of mine at Melbourne University. In 1932 I had fought him in the final of the Melbourne University Lightweight Boxing Championship!

It was left to Expédition Polaires Françaises, in the ship *Commandant Charcot* in 1949, to sail around and chart Sturge Island. The expeditioners also managed to make a substantial landing on tiny Sabrina Island.

Sunday 29 was the most active day of our voyage up until then. The day was fairly calm, with snow showers and a Force 2 SW wind. At 1000 hours, a crew member told McCarthy and me that the whaler was being prepared to attempt a landing. I wondered why Campbell had not informed me. Before there was time to collect equipment or even consider appropriate clothing to wear, Mac and I found ourselves being lowered over the side in the long whaleboat, with oars in our hands. In the boat with us were Campbell, Cook, W/O Homewood, Le Guay, Petty Officer Smith and seamen Wallace, Swinnerton and Tame. The sea was choppy and the air temperature 27.5°F. We rowed in pairs abreast for about $2\frac{1}{2}$ miles towards a small spit of land at the north-western tip of Borradaile Island.

The coast was forbidding in the extreme. The highest point on the island, halfway along the western coast, was later estimated from the ship to be 1,250 feet high. Most of the rest of the island was found to be 500 to 1,000 feet, and along the entire coastline, apart from the north-western tip, high ice cliffs plunged vertically into the sea. This day, however, with cloud at less than 1,000 feet, much of this was not visible to us.

Where we approached, the high ice-cap of the island plunged steeply down to end in rock cliffs leading to a short and narrow boulder-strewn spit with a steeply shelving beach of black volcanic sand. Scattered over the beach like boulders were blocks of ice six feet or so in diameter. A heavy swell broke close to the shore and dashed wash and spray among them.

Cook dropped anchor 30 yards offshore and allowed the whaler to drift in stern first, yard by yard, by paying out rope. When within several yards of the beach, where the surf was dumping, everything seemed to get out of control. Wallace, crouched in the stern, was supposed to jump ashore with a rope and fasten it around a boulder or block of ice. However, when the boat grounded, he remained immobile, apparently hoping for the receding surge to allow him to get ashore relatively dry. But the breaking waves smashed over the whaler's bows, drenching Cook, McCarthy and the others, who sat gripping their oars. Campbell summed up the situation immediately. He grabbed the rope from Wallace and jumped into the surge. Wading ashore, he took a quick turn of the rope around an ice block, but was unable to hold the purchase as

The north-west point of Borradaile Island where Stuart Campbell, Phillip Law and Seaman Wallace landed in February 1948. *POLAR RECORD*, JANUARY 1965

the boat receded in the backwash. Since Wallace still did not move, I pushed past him and followed Campbell. The water was up to my thighs, and filled the rubber knee boots I was wearing.

It was my first experience of immersion in really cold water—around 29°F! The pain was excruciating. I felt as though my legs had been chopped off above the knees. I waded desperately up the sloping sand and tried, with Campbell, to hold the rope anchorage. Wallace then came to life, jumped ashore and joined us just as the rope around the ice block came unstuck. Two large waves crashed over the whaler, slewing it around broadside and causing confusion and panic. The rope caught under the rudder and lifted it off its lower hinge. Another wave surged in. The whaler now had 18 inches of water in her. Cook yelled to us to get aboard again, and he and McCarthy hauled frantically on the anchor rope to drag the boat off and straighten it up. We dropped our rope, scrambled back through the surf up to our waists and hauled ourselves aboard as the others pulled the whaler out from the breakers. By the time we were through the surf line, the boat was half-full of water.

Young Island, Balleny Islands. PHOTO BY L. LE GUAY FOR DEPARTMENT OF INFORMATION

Grabbing our oars, we rowed desperately through the line. Further out, shipping the anchor, we baled out the boat by hand, and re-set the rudder. My legs were freezing in the sea-filled gum boots, and I vowed never again to wear such footwear. What made things worse was that later we were to discover the whaleboat had a hand-operated pump for baling!

We tried to row to a small island to the north-west, about 50 yards long and about five feet above water level. The wind had, however, gone around to the WNW. After striving for some twenty minutes against wind and swell, we had to give up. The oars were heavy and our arms were aching, so we headed back towards the ship, which had been six miles out, but had begun to steam in closer to assist us.

I remember rowing and rowing and rowing, seemingly making no progress and becoming colder and colder, and more and more exhausted, despite the exertion. Snow showers drove across us and curtained off the ship. After about three-quarters of an hour of this, with hands and feet frozen, we reached the *Wyatt Earp* and clambered clumsily but thankfully aboard, up a rope ladder. It was 1330 hours. We had done no better than Captain Freeman, 109 years earlier! Campbell, however, proudly displayed a rock specimen that he had had the presence of mind to grab from the beach as he ran to the boat. It was the *only* geological specimen ever collected from the Balleny Islands!

Snow squalls continued on and off until, at about 1600 hours, the weather began to clear and the ship started a running survey of the eastern coasts of Buckle, Borradaile and Young Islands. Young Island in particular presented a glorious sight as the sun broke through the clouds to shine on its icy slopes. From the sheer coastal cliffs, they rose gently to the plateau summit which our navigator estimated to be 3,250 feet high. Its highest ice cliffs, on the western side, rose to 1,000 feet. The slopes looked marvellous for skiing.

That evening the crew cooked and served the dinner, which was very good. We followed it up with a picture show. After that Mac and I developed cosray film, then fell into bed, exhausted, at 0100 hours. It had been a very full day.

Monday, by contrast, was very quiet and restful. The ship steamed along past the islands, snow fell and, although there was little wind, there was quite a heavy swell. In the afternoon, in thick fog, the wind dropped away and the water became oily smooth. The watch noticed an ice floe upon which was some rock debris. 'Knocker' White climbed overboard onto the floe with a rope around his waist, and picked up some samples. They had almost certainly come from the eastern coast of Buckle Island.

By 1700 hours the fog had lifted between us and Buckle Island, but it was still thick to the north. Heavy clouds hung over the island's summit, but the low sun brightly illuminated the lower slopes through a space in the clouds. The view across the unrippled water produced a placid 'calm after the storm' chiaroscuro of great beauty.

My hopes of making a landing for geomagnetic work were rapidly fading. I had never seen coasts more menacingly forbidding than those we had passed. At the northern end of Buckle Island was Cape Cornish, named after my university acquaintance, a black rock bluff 1,100 feet high, topped with a further 100 feet of ice. Cape McNab, at the southern end, was almost as high, while there were three 600-foot rock bluffs on the eastern side. The whole of the western coast was ice cliffs, as was the rest of the eastern coast. In some places ice falls at the seaward terminations of glaciers tumbled a confusion of mammoth ice blocks one hundred feet into the sea. Our navigator's estimate of the elevation of the island was 3,025 feet.

Young Island, the largest of the group, was almost completely encircled by ice cliffs 300 to 1,000 feet high, while at its northern end was a remarkable rock precipice that rose sheer from the water for 950 feet. It was so utterly vertical that it appeared concave. The surface of the plateau rose smoothly to an elevation of 4,400 feet.

Borradaile Island, almost square in shape, was surrounded by steep rock cliffs capped with heavy crusts of glacier ice, except for a spur at the north-west corner that slopes down to sea level to terminate in the spit upon which we had landed.

Rowe Island was quite small and, totally capped with ice, looked more like an iceberg than an island.

McCarthy stayed up until 0230 hours that night struggling with the 'shower' recorder of the cosmic ray equipment that had given us continuous trouble since the beginning of the voyage.

Tuesday was a rotten day, with a strong SSE wind and heavy snow showers. The gyro compass again gave trouble. The crew engaged in some snowballing on the deck and took the opportunity to scrape up snow in buckets to provide fresh water for their laundering. I did not disturb Mac, who was still sound asleep after breakfast, but worked in the lab until lunchtime. The weather became steadily worse, so we steamed up and down, ESE then WNW, about a mile off-shore in the lee of the giant 1,200-foot bluff at the northern end of Buckle Island, with the wind at Force 7. In between snow showers, at the end of our westerly runs, we could see the open ocean outside our lee shelter whipped into fury by storm winds that must have exceeded Force 10. We were grateful for the shelter afforded by the island because, apart from the discomfort of pitching while hove-to into such a storm, we would have faced considerable danger of icing up the ship, since the air temperature was 25°F and the sea temperature 31°F.

In the late afternoon I stood on the bridge and gazed out, fascinated, at the grim scene. The lowering black rock cliffs rose from the wild seas until, capped with 50 to 100 feet of ice, they became lost in the clouds above. Spindrift from the storm floated past the ship, streaks of light grey on the dark water, while to the north-east, only about a mile away, a line of very thick pack ice was visible. A number of small icebergs could be seen on the radar, dotted around within five miles of us, while the sea about us was

Sabrina Island and the Monolith. PHILLIP LAW

thick with brash ice, the pulverised remnants of ice blocks that had fallen from the Buckle Island glacier cliffs.

Back and forth, back and forth we went, slowly, over a distance of about a mile, with sickening lurching rolls as we turned at the end of each lap. At 1830 hours the wind rose to Force 10, with gusts up to 70 m.p.h.

All night the storm continued, moderating a little next morning but building up locally to Force 9 in the afternoon. Two days of this, at even greater force, out on the open ocean would certainly have been most unpleasant, if not actually dangerous.

Next afternoon the wind began to drop and the weather to clear. We left our sheltered position, since the line of heavy pack ice to the east was coming uncomfortably close. As a forecast of coming fine weather there was a lovely sunset, with gold and crimson tinting a complex pattern of cirrus clouds high above us.

My 35 mm camera, which I had borrowed for the trip, had unfortunately given up; I found that the shutter blind had broken. Unless I could borrow one, my still photography was at an end. I had been taking cine shots on what, today, would seem quite an antideluvian

16-mm movie camera, a Bell and Howell *Eyemo*. Although its speed slowed down in the coldest weather, it still continued to operate.

We were surprised at midnight to see navigation lights. Later, we passed a Japanese whale catcher. On this voyage I was to observe a variety of whales, including a number of giant blue ones. Never again was I to see as many, for the whaling fleets were to decimate them over the next few years. Towards the end of my Antarctic career, in 1966, I was seeing only one or two each voyage.

Thursday 4 March was a superb day—the best of the whole voyage. Campbell lent me the expedition Leica camera which had been given to the Captain, but which he was not interested in using. On developing some of my black-and-white still films I found them over-exposed, so the shutter of my camera must have slowed down in the cold. It was just as well the shutter blind had snapped. With this other camera there was a good chance that my later photos might be more successful.

The sun shone brilliantly from a blue, cloudless sky. There was almost no wind. The ship proceeded to sail a complete circuit of the three islands—Young, Borradaile and Buckle—for a running survey. With regular astrofixes of the ship's position from sextant shots of the sun, with the ship's track accurately plotted, with angles measured to various coastal features, and with photographs and estimated radar distances noted, the survey would enable the RAN Hydrographic Office to draw the first accurate map of the Balleny Islands, with the exception of Sturge, the most southerly. It was encircled with dense pack ice, and we were unable to do more than note the position of its northern extremity.

The circumnavigation of the three islands greatly extended the survey done of the eastern coasts several days earlier, and provided much better accuracy. We completed it at about 1900 hours, off the south-east side of Young Island. Unfortunately for the bathymetric work, the echo sounder had ceased to work. McCarthy took it apart and found that the oscillator, mounted on the bottom of the ship, had frozen up. It was enclosed in a tank of fresh water which had frozen into a solid block of ice. He replaced the fresh water with sea water, which fixed it.

We wandered about the deck in our shirt-sleeves, enjoying the glorious scenery and taking numerous photographs. Humpback and killer whales, and a few blue ones, were plentiful. We saw only one spot where a landing might be possible other than at the place we had so briefly visited. It was a small, shelving beach on Sabrina

Wyatt Earp off Young Island, Balleny Islands, 1948. PHOTO BY L. LE GUAY FOR DEPARTMENT OF INFORMATION

Islet, with a steep snow slope behind. Numerous Adélie penguins cluttered the slopes. I believe this was the place where, some years later, members of Expéditions Polaires Françaises landed.

Later in the afternoon I saw a remarkable phenomenon. At a distance of about four miles appeared a large castle-shaped turreted berg, some 100 to 150 feet in height. Every few minutes a great sheet of water and spray was flung high in the air, higher than the berg itself. On such a calm windless day, this was hard to explain. All I could think was that there might be a cavern in the iceberg to which the sea had entry. A swell surging in could be channelled into a narrower and narrower passage, terminating in a vertical cliff which, acting as a kind of wedge-shaped blow-hole, could produce the effect I observed. It still appeared a remarkable magnification of what could have only been a slight swell.

Our fine weather was short-lived. On the Saturday, the ship heaved herself all over the place as we headed westward again, for another try at the Continent. Commonwealth Bay lay some 500 miles on, and I was glad to be heading there again. The Balleny exercise was finished.

We had spent all Friday stooging around in the lee of Buckle Island again; this was pretty boring. Loewe decided to try a bathythermograph run. This instrument is something like the hydroplane used in World War II to drag behind a ship to cut the moorings of mines. It is trailed out astern and, depending on the setting of its hydroplanes and the speed of the ship, descends to a certain depth. Temperatures of the water traversed are recorded as a trace by a stylus, on a small strip of smoked glass. This was the first time we had tried out this apparatus. The trace we obtained was poor, but at least we found out how the thing worked. It was extremely hard work winding it in on the hand windlass against the forward speed of the ship.

Sunday was a better day. The ship stopped for four hours to allow the Chief Engineer to carry out maintenance work on the engines and to change the engine oil. We did another water bottle haul, with collection of samples and temperatures from 200, 150, 100, 50, 30 and 15 fathoms. This was carried out on the after boat deck in a heavy snow storm. As the ship drifted across the line of the submerged wire, the hauling-in procedure was even more difficult than usual. Loewe and I had no other assistance, and found that 200 fathoms of wire took a lot of effort to haul in.

At night the cook was off duty and it was my turn to perform, assisted by Campbell and McCarthy. I had consulted the steward about the kinds of rations available in the store, and had designed what I thought would be an interesting dinner, as different as possible from our usual fare. Cooking it was, however, more of a problem than I had anticipated. The wind was Force 5 from the ENE and the ship was rolling in its usual fashion. It was almost unbearably hot in the galley. The menu I had planned was:

Soup	Crême de Balleny
Entrée	Omelette à la Young
Dinner	Macaroni Borradaile with sausage roll and Sauce Penguine
Sweets	Crêpes de pommes à la citron
	Coffee
	Cheese

The soup proved straightforward. The omelette, a mix of egg powder, powdered milk, bacon, onions, herbs and pickles, was put in the oven to cook in two large flat baking dishes, but the rolling of the ship spilled half the contents out onto the steel trays of the oven. There they cooked just the same! I scraped up the product and successfully served it. The sausage rolls I had baked earlier, and they were heated up in the oven without difficulty. Most of the apple fritter mixture, however, finished up on the floor of the galley as a result of various lurches of the ship as I was handling the large pan. Campbell and McCarthy worked feverishly as my assistants and, one way or another, we served a creditable meal to the ship's thirty-six diners. Thank God the others from the wardroom washed up!

Afterwards, sweating profusely, I went to the lab and had a hot sponge bath with a quart can of water that I had saved for two days from my ration.

The water shortage, next to the motion of the ship, was the most unpleasant part of the voyage. For most of this second journey we had been rationed to half a jug (one pint) of water per day per person, for all purposes other than drinking or developing scientific film. This particular day the situation became particularly irritating. We were told that the day's consumption had totalled 190 gallons. A few days earlier, we were told one ton of water had been used. These figures seemed absurd in the light of the allotted ration! I reckoned that either the record of the consumption was wrong, or else the fresh water tank was leaking—or some people were cheating. In any case, the order was given to lock the bathroom door at night. Next day a further rule was introduced—fresh water for all purposes was to be drawn from the galley tap, so that the cook could keep some sort of check on what was going on.

Whether as a result of this, or for other reasons, the day's consumption moved down to a respectable 90 gallons. I still couldn't see how thirty-six men, even cheating, could have used the 190 gallons previously listed!

With adequate forethought, the refit of the ship should have provided for a supply of hot *sea* water for the washrooms. I wondered whether the fresh water tanks had been completely filled on leaving port. There had been some talk of decreasing the contents to reduce the weight of the overloaded ship, but whether this had been done or not I was unable to find out. I thought, too, that we might have been provided with special soap for use with sea water. This would certainly have helped!

Some days later we were instructed to draw all water for washing from the snow melting tank. However, since the crew had been putting into it the snow they swept up from the decks, this was a crazy idea: it produced water almost as salty as sea water, only much filthier. If clean snow, newly fallen, had been collected instead of being used for snowball fights, this would have made more sense. Further, we had had plenty of time, one way or another, to gather ice from ice floes or bergy bits for the snow melter, but we had only done this once. Then the product had soon been used up in a spate of laundering. All in all, there had been no coordinated attack on the water problem.

In my later ships the situation was greatly improved because the vessels were fitted with evaporators to generate fresh water from sea water. In the *Kista Dan*, however, even this was not adequate to supply more than enough for carefully monitored use of this valuable commodity. Regular recourse was made on that voyage to collecting ice for melting while in the pack ice zone.

I remember coming home at the end of the *Wyatt Earp* voyage and experiencing the intense pleasure of turning on taps in my house to obtain unlimited fresh water. I still, after all these years, regard the availability of fresh water in our normal urban existence as one of the greatest benefits of civilisation. I have continued to be unable to use water in any but frugal fashion.

On 10 March, Campbell commented in his diary on the inadequacies of the *Wyatt Earp*:

The *Wyatt Earp* is not a very good ice ship; she is too light and has too blunt a bow to crack anything but the very lightest ice. When we hit even a small old floe we take a very solid jar and it is doubtful if she could stand much of that sort of work. She does not manoeuvre very well either and is awkward in turning to starboard.

I am quite convinced now that this ship is far too cramped and crowded for a long voyage. It cannot comfortably take more than its own crew, as run at present, with perhaps three passengers. It is impossible to run this as a Naval ship; there are too many people on board with too little to do. We should reduce the crew by about five and make everyone work watches.

9

The Last Approach

THE MORNING OF Friday 12 March saw us heading south towards the region of the Mertz Glacier Tongue. At 0700 hours we passed through a line of icebergs which were obviously grounded, for the depth decreased rapidly to eight fathoms as we crossed the line, and then increased again. Apparently the Captain could see nothing promising ahead, for we changed course to the east before reaching the pack.

After lunch we saw two Crabeater seals on an ice floe. Someone decided that they should be shot for meat. John Homewood and one of the sailors, using .303 rifles, fired at them from the deck. John killed his target, but the sailor only wounded his, and it slid off into the sea. A number of those in the wardroom objected to the whole business. 'Knocker' White, in particular, became quite sulky about it.

The dead seal was hoisted on board. I skinned it and one of the seamen cut out the liver and sliced steaks off its back. I cut out a patch of skin about three feet by two from its back and spent an hour flensing $1\frac{1}{2}$ inches of blubber from it. Every tiny bit had to be scraped off. It was very cold work. I then nailed the strip to the external bulkhead of my laboratory to dry over the next week or so, for I intended tanning it when I got home. The hair was a lovely pale golden colour, typical of the 'fur' of a Crabeater seal in the late summer after it has been bleached by the sun, before the old coat moults.

Years later, I would present a piece of this golden-haired skin to Lady Mawson, who wanted it to make a special cover for a book she planned to give to a friend.

That night we had the steak for dinner. It was a welcome change from the tinned meat of the voyage so far—something to get one's teeth into. It was my first taste of seal meat. I was surprised that it did not taste 'fishy'. It was, in fact, rather tasteless. The flesh was dark red when raw, like kangaroo meat, and a deep purplish black when cooked. It was stringy to chew, but not tough. Everyone seemed to enjoyed it.

The flensing had also left my hands gloriously clean and soft.

About 6 p.m. we reached the region of the beautiful domed icebergs off the Ninnis Glacier Tongue. I nostalgically remembered our departure by moonlight seventeen days earlier, and hoped that the next day would be fine.

Saturday did dawn fine. The Captain decided to send up the plane. The ship sheltered behind a huge domed iceberg, many miles long and over 300 feet high, to reduce the swell. What followed was a bit of a circus. Bosun Homewood knew his stuff: he wanted to unload the aircraft first and, after that, to take off the motorboat. But Cook decided that the motorboat should be moved first. So the plane was first hoisted above the deck into a suitable position, after a couple of hours had been spent rigging control guys and other tackle.

The attempt to unload the motorboat was unsuccessful. Its hull was stoved in against a stanchion. The only alternative was to hoist out one of the whaleboats to act as a tender and rescue boat. This meant further rigging, and it was past lunch before this was accomplished. I reflected that I should hate to have to launch the whaler in a hurry, if a man fell overboard!

There was little room for manoeuvre when unloading the aircraft. With the ship on an even keel and the boom swung out, the clearance from the front of the floats to the side of the ship was only a few inches. In practice it was a bit more, for the weight of the aircraft hanging from the end of the boom caused the ship to heel over.

When it came to hoisting the aircraft, everything went smoothly. Gray took off for a test flight, with his mechanic 'Spike' Jones as passenger. The whaler, with Le Guay aboard to take photos, rowed hither and thither. The rest of us leaned over the rail and watched.

Taking upper wind balloon measurements onboard *Wyatt Earp*, from left: Edward McCarthy, Fritz Loewe and Phillip Law. PHOTO BY L. LE GUAY

After a short flight the plane landed. Exchanging Jones for Le Guay, it took off again. The plane-to-ship radio communication worked well. We learned that Gray could not see the Continent. He reported a thick mass of bergs, choked with pack ice, for at least forty miles south, but clear water for about twenty miles south on each side of this accumulation.

We were, however, destined not to take advantage of this information. The Captain had decided it was too late in the season to attempt any further 'southing', and to run east for a short time, taking soundings, before turning north for Macquarie Island.

When the aircraft landed again at 1530 hours we all helped hoist it on board by hauling on various control guys. Then the whaleboat was brought aboard. Apparently it was decided that there was to be no more flying, for the ship turned north. We began to help the airmen dismantle the plane ready for stowing for the return voyage. It started to snow, and it took about three hours in a biting wind to remove the large float, turn the plane around and set it down on its deck cradle. We got to dinner at 1930 hours, very cold and tired.

Campbell again wrote about the Kingfisher:

> Today's effort convinced me that I'd made a mistake in selecting this aircraft for the job. It is too awkward to handle with the gear on this ship, and takes up too much deck room. Had it been stowed, fully assembled, as I originally intended, I think it would have been better, but still too big for us to handle comfortably. Used as an aircraft for long reconnaissance flights occasionally, the trouble of handling does not matter, but these days, it is not good to use a single engined aircraft for long flights in the Antarctic, except for stunt purposes, and we want a light aircraft that can be used merely for increasing the range of vision from the ship. It must be carried in one piece ready to fly, it wants no wireless or other gadgets, and should be able to be put over the side in half an hour, fly for half an hour and be re-stowed in half an hour. Then it would *really* be an adjunct to the ship's work. As it is with the Kingfisher, it is rather a nuisance having it on board—it is too much of an operation to fly it.
>
> For a tender, we want a small handy 14 or 16 foot skiff with outboard or other small motor, and nothing more. Everything must be as light and simple as possible.

The Captain began an easterly run to record depth soundings along a track roughly parallel with the Ninnis Tongue–Cape Freshfield coast. The sun came out early next morning, but it soon clouded over again. The ship stopped for a short time to allow the crew to tack further sheets of canvas to the hull on the starboard side, where a leak into the wardroom had again developed. While this was going on, Loewe took bathythermograph readings and was amazed when the ship started to move again without anyone informing him. If water bottles had been hanging out at the time they would probably have broken away.

This was just another example of the lack of consideration for the scientific program during the voyage. Earlier in the voyage, Cook had shifted the barometer from the drying room to the after deck, without consulting Loewe. As Cook did not know the precautions necessary in carrying a Fortin barometer, it was probable

Launching the Kingfisher aircraft. PHOTO BY L. LE GUAY FOR DEPARTMENT OF INFORMATION

that the calibration of the instrument was changed; Loewe had been given no opportunity to carry out the usual precaution of taking readings immediately before and after the transfer. The general attitude of the non-scientists was: 'What good are these results? Why do you want to do that? Why expend all that effort? Why waste time pulling up water samples? Who would want to know anything about an Emperor penguin's egg?'

I have mentioned the resentment of the non-scientists in the wardroom to the fact that McCarthy and I had been relieved of our watchkeeping duties. Bond and White had been particularly annoyed. Before the aircraft was flown, Loewe had asked to be allowed to fit a thermograph to the wing. The instrument had been prepared for the flight and would have taken only five minutes to attach. The Captain said it was not worthwhile, and would not allow it. Yet it took four hours of messing around before the aircraft was hoisted outboard! Loewe was so annoyed that, brooding over the matter, he inadvertently allowed the balloon theodolite to tip over, damaging it so that it could not be used again for the voyage.

An entry in my diary reads:

> We scientists have been regarded with amused tolerance—in some cases resentful intolerance—as rather queer examples of a rare and foolish species. The sadistically cruel way they speak of Loewe behind his back or tease him to his face reminds me of adolescents at school. 'Knocker' White, for example, can turn on a fit of sulks as a protest against the killing of two seals, but can maliciously torment Loewe, a most learned, tolerant and kindly person, whose personal idiosyncrasies make him an easy butt for schoolboy wit.
>
> At 1100 hours on the Monday we passed a giant iceberg, some fifteen miles long, together with many domed bergs. The latter appeared to be grounded because, as we approached, the depth decreased rapidly.
>
> It is, I think, too much to expect that the Navy will make any attempt to analyse and write up the information obtained during the voyage on the interesting bottom features we have sailed over—moraines, canyons, iceberg-planed flats, depths at which icebergs have grounded, and so on. The only result will probably be a line of soundings on a hydrographic chart.

By the Tuesday we were heading north for Macquarie Island. Loewe and I did a bathythermograph haul and obtained an interesting trace with a kink halfway along it, indicating a current of colder water between depths of twenty and thirty fathoms.

Petty Officer West informed me that the Captain had told him I would be allowed no more than one bucket of water a week for developing our cosmic ray films and general maintenance of the batteries and the laboratory in general. I tackled Oom who, while not exactly denying it, looked surprised and said there would be no limit—that we would be given all we needed. He said he expected to have plenty of water after he had restocked at Macquarie Island from LST 3501. I guessed what probably had happened was that he had said off-handedly to West, when discussing the distribution of water on the ship, that one bucket should probably be enough for the cosray lab. West had then taken this as an instruction.

The temperature of the air had risen, and ice on the rigging and elsewhere had thawed out. As cylinders of ice that had coated ropes aloft fell from the rigging, we gathered them to help with the fresh water supply. While I was stooping to this job, Gray shook the ropes to loosen further material. A great bar of ice the size and shape of a marlinspike fell forty feet and struck me on the head, almost knocking me out. If I had not been wearing my padded ANARE cap, I could have suffered serious injury.

Water restrictions were relaxed, however, to the extent of allowing us half a can (about a quart) of fresh water daily from the galley.

The Captain offered to stop the ship twice during the day for Loewe's bathythermograph work, much to Fritz's surprise. On each occasion we obtained evidence of the cold water layer at the twenty to thirty fathom depth. For the last thirty hours, the weather had remained calm and fine, but in the afternoon a westerly wind sprang up, and the ship began to roll heavily.

Thursday was a fine, sparkling, sunny day, with a fresh westerly, a large swell, flying cumulus clouds and an air temperature of 37°–38°F. We expected to reach Macquarie Island on the Saturday. That night was the worst I had spent on the voyage. With a SE wind Force 6 to 7 and a beam sea, the ship rolled abominably, nearly tossing me out of my bunk on several occasions. I did not get to sleep until 0500 hours. Campbell also recorded that night as his worst.

Next morning the wind swung to the NNE, at Force 6. The rolling during the night had flooded the wardroom again, and some of the starboard cabins. Engine trouble with the fuel pump stopped the ship for four hours. The crew trailed a sea anchor and hoisted the fore trysail to stabilise the ship and reduce the rolling. Campbell wrote:

> At one stage, struggling to drink a cup of tea in the wardroom, I heard a terrific crash from the galley and, looking in, saw the cook on the deck sitting in the contents of the slop bucket with a dixie of rice pudding, on its side on the bench above, pouring down on his head.

During the day I found that the non-naval personnel—Campbell, Bond, Le Guay, Gray and Loewe—had given up watchkeeping! After all the fuss earlier in the voyage, I found this wryly amusing.

That night the wind swung around to the south-west and blew at Force 8, with gusts of up to 10. As we were sailing on course 020° the great swells arising from the storm swept up astern and drove us along at what was an all-time record speed just on ten knots.

The spoon-shaped stern was not, however, designed for such seas. Time after time, a giant swell would crash over the stern, driving a wall of water through the badly fitting after doors along the passageways on both sides of the ship. Those on watch were continuously employed bailing out water. They were joined by those whose cabins were being flooded. I thought we had experienced everything possible on this voyage—but while working in my lab, several heavy rolls threw me from one side of the bulkhead to the other, severely bruising my elbows and buttocks.

10

Macquarie Island

AT 0535 HOURS on Saturday 20 March the watch sighted the Bishop and Clerk Islands, the southerly offliers to Macquarie Island. At 1230 we anchored in Buckles Bay, in about twelve fathoms of water. The wind backed to the NW and decreased to Force 4. The air temperature was up to between 40° and 44°F. Nestled in the lee of Wireless Hill was the ungainly outline of LST 3510, commanded by Lieutenant Commander George Dixon, who had been given the task of setting up the Heard Island and Macquarie Island stations.

This was my first view of Macquarie, an island I was to visit often over the following years. The verdant green of the grassy escarpment slopes contrasted strikingly with the rock and ice of the Ballenys. It was amazingly beautiful to our eyes, which had seen no vegetation since leaving Australia. The general height of the island is about 600 feet and, from this plateau, hills rise at various points to between 1,000 and 1,400 feet, with Mount Hamilton near the southern end at an altitude of 1,420 feet. At the northern end of the island the plateau drops down almost to sea level, where a long flat isthmus connects to another high point, Wireless Hill (358 feet), at the extremity.

In December 1911, Douglas Mawson's Australasian Antarctic Expedition (AAE) had landed to set up a scientific station on the

Wyatt Earp off Macquarie Island. PHOTO BY L. LE GUAY FOR DEPARTMENT OF INFORMATION

isthmus, hard against the foot of Wireless Hill. This received its name because the AAE had erected wireless masts for communication with Australia on the top of the hill. The ANARE station was to be established on the old site of the AAE station, of which only wooden ruins remained. It had been abandoned at the end of 1915, when the relief ship sent to evacuate the 1914 station team had disappeared without trace during a violent storm.

Immediately after lunch a naval barge, LCVP (Landing Craft Vehicles, Personnel), came over from the LST and picked up Campbell, Loewe, McCarthy, Bond, Le Guay, myself and some of the crew and took us back to the LST to indulge in the luxury of a shower. We found that a boat was about to go ashore, so I postponed my wash and joined it.

We went in by LCVP to a buoy just beyond the surf line in Garden Cove, transferred to a dinghy, and hauled ourselves to the beach along a grass line stretching from the buoy to a rock on shore. There were several anxious moments as we crashed through the heavy surf. It was a very rough landing place, with ugly black heads of rock sticking up through the sea, and long tendrils of giant kelp covering the surface. In the end, however, we got ashore, safe and dry.

From the black sand of the small beach a D4 Caterpillar tractor was towing heavy cargo sledges, loaded with fuel drums, back and forth to the station site. It moved through a muddy quagmire churned up by the traffic of bulldozer and Army DUKWs, amphibious six-wheeled motor vehicles. On the beach lay the shattered remnants of a LCVP and several wooden landing pontoons whose rigidity had proved unequal to combat with the waves, and which had been pounded to pieces on the rocks. Luckily the expedition had been provided with the DUKWs, which coped well with the difficult conditions. But never again did the ANARE use the Garden Cove landing beach. In subsequent years DUKWs came ashore at the beach half a mile south, in Buckles Bay.

The LST had arrived on 7 March and the station was already beginning to take shape. The cook hut had been erected and was serving also as a mess hut. Nearby were tents furnished with stretcher camp beds, serving as dormitories for men whose tasks ashore made it inconvenient for them to return each night to the LST. Huge piles of stores covered by tarpaulins lay stacked on wooden railway sleepers.

Gentoo penguins were numerous along the beaches and could be seen 'porpoising' and swimming underwater around the ship. My cosmic ray colleague in the MI party, Charles Speedy, took me to the cliffs above Garden Cove to see Rockhopper penguins, and to Hasselborough Bay to see my first sea elephants. Those huge seals were quite repulsive, lying like slugs on the sand or wallowing in stinking muddy pools in amongst the grass tussocks of the isthmus. Giant petrels, huge grey and white birds that reminded me of vultures, gorged themselves disgustingly on the carcases of dead elephant seals, while brown skuas, about the size of crows, scavenged amongst the penguins and displayed their aerobatic skills in the gusty winds. Low heavy clouds obliterated the plateau, and rain fell incessantly. The light was too bad for photography so, wet and cold, I returned to the beach and took the next boat out to the LST to have my belated shower. Upon arrival at the ship's side, however, I was told that the Captain wanted all the *Wyatt Earp* people back on board immediately, since the wind was rising and he wanted to put to sea rather than remain at anchor overnight. So again I missed my bath!

Next day, Sunday 21, was one of intermittent sunshine. I spent it taking still and cine photographs of the men's activities ashore, and of the wildlife. Two bull sea elephants fought a challenge match on the back beach, rearing up on their flippers and slashing at each other with their tusks as they crashed their bulk against each other's

On Macquarie Island. *Labuan* (left) and *Wyatt Earp* (right) are visible out to sea. PHOTO BY L. LE GUAY FOR DEPARTMENT OF INFORMATION

chests. It looked grand through the viewfinder of my cine camera and I was elated to have captured the sequence. I was not to realise how commonplace this scene was. Leaving them to their contest, I went out to the LST. There I radioed the *Wyatt Earp* for permission to remain overnight. Then I finally caught up on my long-awaited shower. And how marvellous it was!

That afternoon a slapstick comedy was enacted by the ship-to-shore transport system. At about 1600 hours an LCVP left the LST and proceeded to the shore to pick up crew members and ANARE men, to return them to the LST and the *Wyatt Earp*. Ashore was the LST's bosun, a Cockney Englishman named Alf Hayter, who was a Navy veteran from World War II and a highly efficient seaman. He ruled the sailors of the LST with the iron hand of a regimental sergeant-major and a flow of invective that I have never heard equalled.

The LCVP picked up a full load of passengers and set out for the LST. Giant kelp, however, became twisted around its propeller, and it began to drift helplessly before the wind. The LST doctor, Surgeon-Lieutenant Peter Blaxland, with two sailors, launched a dinghy from the shore carrying Alf Hayter on a rescue mission. They rowed to a nearby head of rock that stuck up from the sea about thirty yards on the shore side of the LCVP. Hayter jumped from the dinghy onto this 'island', and called to the dinghy for a line. A sailor cast him a light heaving line, which was useless for towing. Hayter had to call for a heavy 'grass' line, which was duly attached to the heaving line at the dinghy, and hauled to the rock. Hayter then fastened this line to the rock while the two sailors rowed across to the LCVP with the other end of the rope. With this duly made fast to the LCVP, the danger of it drifting either out to sea or onto the shore rocks was removed. When the dingy tried to go back to fetch Hayter, however, the men, rowing into the strong wind, could make little progress. At last they came within ten yards of the rock. By then the wind had risen even further, and heavy seas were breaking over it. A closer approach was deemed too dangerous.

One of the sailors called out: 'Jump!' His idea seemed to be that Hayter should swim out to the boat. His language in reply, according to the sailor's later report, exceeded any of his previous efforts! The exhausted sailors could do no more, and the dinghy drifted away with the off-shore wind, and out to sea around the northern tip of the island.

The whole event had been viewed from the LST through binoculars by Lieutenant John Lavette. He sent a radio message to shore requesting a DUKW to come to the rescue. The DUKW, however, was unloading a cargo of fuel drums, and it was almost half an hour later before it left Garden Cove. It first had to head out to sea to rescue the drifting dinghy, which took a considerable time. It then took a line to the LCVP. The ANARE cook, who cast the line, was dragged overboard because he hung onto the rope and tried to hold the tension between the two craft instead of taking a turn around a bollard. He was duly hoisted back onto the DUKW, which then towed the LCVP out to the LST.

By this time it had grown dark. Hayter was still on the rock. The tide had risen, the rock was half-submerged, and he was drenched each time waves broke over it.

At long last the DUKW, its headlight casting a bright beam through the darkness, sailed from the LST to the rock. It backed in close—but not close enough. Hayter had to jump into the water and swim the intervening few yards. Some seconds later he was hauled aboard the DUKW. It was at that point that, directed at sailors in general, Hayter's language eclipsed all his previous records.

11

The Return

IT IS ASTONISHING how quickly one regains full strength and vigour on returning ashore after a voyage. Although I had long since got my 'sea legs', and could stand even the most violent motions of the *Wyatt Earp* without becoming ill, my lethargy and general 'liveryness' during stormy periods persisted. Ashore at Macquarie Island I was a new man—alert and energetic, with a huge appetite and able to call on completely different, deeper sleep. I enjoyed the change immensely.

On Sunday 21, at 2200 hours, the ship's cat Mimi (alias Itsy Bitsy) gave birth to two kittens. This surprised us all because she had been such a tiny thing when she came aboard at Hobart.

On the Monday the *Wyatt Earp* took on 18 tons of water through a line slung between the ship and the LST. A lot of messing around was involved. The sailors on the LST were just inexperienced kids. Poor Alf Hayter nearly had apoplexy as he screamed sarcastic abuse at their incompetent efforts.

Macquarie Island has its own peculiar brand of weather. A completely fine day is extremely rare. What passes generally as 'fine weather' consists of a wild westerly wind of 25 to 40 knots, and scudding clouds and intermittent bursts of sunshine alternating with showers of rain, hail or snow. Rarely are the Macquarie plateau and the summits of its hills visible for a whole day at a time—even in the finest weather. In bad weather the wind screams, the smog comes down almost to the level of the isthmus, and the rain drives horizontally. The giant waves of the Southern Ocean pound on the shores and the salt spray whips right across the narrow isthmus. In the greatest storms, the sea has been known to wash across the tussocks from Buckles Bay to Hasselborough Bay.

We left Macquarie Island with its small ANARE pioneer party of fourteen men at 0608 hours on Wednesday 24 March, rounded the northern tip of the island, and set out for Australia on course 300° into a Force 9 SW storm. Was it rough!

Here is Cook's description:

> Macquarie Island is the visible manifestation (together with the Bishop and Clerk Islands to the south and the Judge and Clerk Islands to the north) of a submarine mountain chain which rises very steeply from depths of thousands of fathoms. The long unhindered westerly swells of the Southern Ocean build up alarmingly over this shallow shelf and the seas in a gale must be seen to be believed. Once having committed ourselves it was safer to continue on rather than attempt to turn and the next two hours were in my estimation the most dangerous of the whole trip. We crossed the mile-wide shelf at a speed of one knot with engine revolutions on for seven. It was fantastic that the old ship could live through such water without taking dangerous amounts on board. We rolled like a log, pitched, tossed, yawed, rose to incredible heights on crests and plunged to abysmal depths in the troughs. She did everything but stand vertical although once or twice she tried her damnedest. Most of the time one felt like a buckjump rider expecting to be pitched off or to feel the mount fall under you.

My diary records:

> The seas were as high as any we had seen. The *Wyatt Earp* would climb to the towering crest of one, then plunge with sickening velocity—faster than the fastest lift—to the trough, where she would bury her bows in the next swell, before shaking free of the engulfing water and laboriously climbing the steep wall ahead. The accelerations on the downward path would be greater than 'g' (the acceleration due to gravity), for the upthrust of the passing wave under the stern would tip the ship and propel its bows downwards so that one's feet would leave the deck. (In bed at night in such waves the bunk falls away from beneath one's body and one catches up with it with a jolt at the bottom of the drop.) The ship made a maximum speed of only $2\frac{1}{2}$ knots throughout that day.

During the morning I had a very lucky escape from serious accident, or even death. I had gone up the companionway to the boat deck on the way to the laboratory, wearing my rubber knee-length gum boots because the decks were awash. At the top of the stairs

I waited for the roll of the ship to present an up-hill slope, then hurried up it to grasp a rope hanging from the boom of the after trysail to secure myself against the 'throw' of the ship as she rolled to the other side. When this happened my rubber soles slipped on the wet deck, the rope was torn from my grasp, and I was hurled literally horizontally through the air. Fortunately I struck the railings around the top of the companionway, otherwise I would have gone feet-first overboard, with no-one around to notice! As it was I suffered severe bruising to the left thigh which took the brunt of the impact. I was lucky not to break my femur.

The 27th was Easter Saturday, and our fifth really sunny day for the voyage. The wind was from the north, gradually freshening, and sea and air temperature were a balmy 50°F. This was not to last, however. The radio from Australia gave us a storm warning and we ran into this storm next day. Not that it was anything special—but again the seas were huge. That night I stood on the bridge with John Homewood and watched the giant waves sweeping up on our port quarter. The moon filtered through scudding clouds and the ship rolled both gunwales under, with the deck awash forward of the bridge. One huge wave began to break twenty yards away and came screaming down upon us at about the level of the bridge. I thought it would take us green, but the tough old ship threw sheets of water to each side as the bows lifted, and rode nonchalantly over the crest. It set us rolling, however, and we performed several rolls of more than 40° before evening out again.

Over the last two days the sea calmed and we had a very good run. I found I was developing a carbuncle on the back of my neck. This followed a sty on my eye and an infected nostril. I concluded I must be pretty run down in general health.

The *Wyatt Earp* passed through the Heads into Port Phillip Bay at about 1730 hours on Wednesday 31 March. To our dismay, we proceeded westwards to anchor off St Leonards on the Bellarine Peninsula, east of Geelong. Upon enquiry I learned that the Navy had made arrangements for an official welcome at Port Melbourne at 0900 hours the next morning. The good run of the last few days had put us ahead of our estimated time of arrival, and the arrangements made could not be altered.

All of us were very disappointed. After such a long absence we had been counting the hours until our arrival and this overnight stay at anchor within sight of shore lights was hard to take. In fact there was almost a mutiny among a few men, who kicked around a proposal to take one of the ship's boats and row ashore. Instead, a party of considerable proportions developed, continuing all night. Almost everyone got drunk. At about 0300 hours, we noticed Tony Bond was missing. Several of us went down to his cabin to see if he had hit his bunk. He had, and was sound asleep, uncovered and wearing only jockey shorts. Thereupon was perpetrated the most diabolical practical 'joke' I have ever witnessed. Le Guay extracted Tony's limp penis from the fly of his shorts and put a loop of string over the end of it. He attached the other end of the string to Tony's wrist, without disturbing his slumber. Then, as we watched, he tickled Tony's nose with a piece of cotton. The effect was dramatic. Tony swung his hand to brush his nose, the string tightened and Tony leapt three feet out of his bunk.

At 0656 hours the ship got under way, and steamed up the bay to her berth at Station Pier, Port Melbourne. The Captain set us all scrubbing the boat deck and after deck. After the night before we were still under the weather, and I think this was designed to sober us up.

At 0715 hours, Campbell, who had been at work on the boat deck, decided to move down to the after deck. Instead of going down the companionway, he decided upon a short cut. He climbed through the rail and lent out aft, intending to grasp a halyard which ran from the end of the trysail boom down to a point just starboard of the rudder post. He missed—and went straight overboard.

Cook told me later he had always had a small ambition to have the opportunity of calling 'Man overboard'. He saw Campbell go, but was so astonished (or so drunk) that he was left gasping, trying to get the words out. A sailor beat him to it.

The ship turned about and manoeuvred beside Campbell so that he could climb up a rope ladder that had been dropped over the starboard side. Le Guay obtained a lovely photo of Campbell climbing the ladder looking like a half-drowned rat, his cotton clothing clinging to his dripping body.

Our arrival at 0910 hours at station Pier, Port Melbourne, was an anticlimax. There were very few people at the wharf, and almost none of the press. Apart from wives and friends, there was a small group of Navy brass that accompanied the Captain to his stateroom. The non-naval personnel joined their friends and relatives, and made a rapid departure.

I was met at the ship by my wife Nel, with two of our friends, Alan and Rana Plowman. They drove me home and invited Nel

Right: 'Man overboard': rescuing Stuart Campbell after the incident in Corio Bay, March 1948. PHOTO BY L. LE GUAY

Below: *Wyatt Earp*: the end of a gallant little ship.

and me to dinner that night at their home. The reunion with my wife and the luxury of a shower with unlimited hot water set up what was to be a most memorable evening.

The dinner was superb—in extreme contrast to the fare I had been having for the last three months. Alan had just returned from pig shooting in the Riverina and Rana had roasted a young wild pig and accompanied it with a delectable array of succulent dishes. The food, the wine and the company provided the perfect contrast to my spartan existence on the *Wyatt Earp*.

Next day I returned to the ship for my gear. As I left I looked back at the ugly, tiny, wooden vessel upon which I had spent three and a half months. I felt a ridiculous twinge of sadness. The ship, despite her discomfort and shortcomings, had become home, and I had developed a strange sort of affection for her. I knew I would never sail on her again.

What happened to the *Wyatt Earp*? She was paid off at Melbourne on 30 June 1948, and sold to the Pucker Shipping Company (Victoria) for 11,000 pounds. The steel sheathing was stripped off and the long-distance fuel tanks were removed to increase the hold capacity. Renamed *Wongala*, she was used as a 'tramp' ship between Victorian and Tasmanian ports.

In 1956 she was bought by the Ulverstone Shipping Co. and renamed *Natone*, presumably after the small town Natone near Ulverstone, Tasmania. After some eighteen months trading in Tasmanian waters, she began service between Queensland ports, with an occasional voyage to Victoria.

In January 1959, *en route* from Cairns to Brisbane, she encountered two severe storms. She successfully weathered the first but, during the second, sprang a leak. The pumps could not cope with the intake of water, which flooded the engine room. Sails were rigged, and she managed to reach Rainbow Bay. There her moorings failed to hold her, and she began to drift. During the night of 23–24 January 1959, she ran aground near Mudlow Rocks, some six miles north of Double Island Point lighthouse. Her crew of eighteen reached the shore using her hatch covers as rafts. She broke up within a few weeks.*

* From information kindly supplied by the Navy Office, Department of Defence, Canberra

12

AFTER THE *WYATT EARP*

AFTER RETURNING FROM Antarctica on the *Wyatt Earp* I took my cosmic ray equipment on a return voyage from Sydney to Japan to extend my investigations into the latitude variations of cosmic rays to encompass tropical latitudes.

At the end of 1948, the equipment at the island stations was returned to Melbourne. In 1949, Professor Leslie Martin, of the University of Melbourne Physics Department, decided not to continue in the cosmic ray field, and I made arrangements with Professor Lester McAulay, of the University of Tasmania Physics Department, to take over the work.

A rebuilt set of equipment was sent to Macquarie Island in 1950, where it operated for two years. During 1952, 1953 and 1954, the apparatus was redesigned in Hobart for Antarctica. In 1955 it was installed at Mawson. Continuous cosmic ray recordings have been made there from that time right up to the present. This has largely been a result of the continuing enthusiasm of Dr Geoffrey Fenton, Dr Peter Fenton and Dr Robert Jacklyn at the University of Tasmania.

On 31 July 1948, my secondment to the Antarctic Division had ceased. On 1 August I returned to my position as Lecturer in Physics at Melbourne University.

My future at that time was uncertain. I held a tenured position at the University, but my salary was low. Any prospect of promotion was quite distant, depending as it did upon death or retirement creating a vacancy higher up.

I had had a taste of the Antarctic and of a life that combined my love of adventure with my interest in science. I was keen to continue such a career if at all possible. The situation in the Antarctic Division was, however, obscure. There was no guarantee that the government would continue to support programs at Heard Island and Macquarie Island if the main object of the political exercise —strengthening our claim to Australian Antarctic Territory by establishing a permanent base on the Continent—could not be implemented. The *Wyatt Earp* had proved quite unsatisfactory as a polar vessel, and there was no immediate prospect of finding a replacement for her.

There was a rumour at the time that Campbell would not remain with the Division. His interests were adventure, not science, and the Antarctic Continent, not the two islands. I believed that, if he could see little prospect of landing an expedition to establish an Antarctic station within the next couple of years, he would return to his permanent position with the Department of Civil Aviation.

My goals were different. Certainly, I was as keen as Campbell to set up an Antarctic base, but I would have been quite content to operate the Heard and Macquarie stations and develop their fascinating scientific programs while waiting patiently for the opportunity to build or charter a polar ship.

There was a further uncertainty. My return to the University had left the Antarctic Division without a scientific co-ordinator. Unless the whole show were to be closed down, someone would have to be appointed to fill this gap.

Although our personalities differed fundamentally and I did not admire Campbell's style of leadership, I had found it possible to work amicably with him. I was quite prepared to accept a permanent position as Senior Scientist if one were to be created. But what did the Department of External Affairs plan to do in this regard? I did not know. I asked Campbell to keep me informed of any developments, stating explicitly that I would be an applicant should any such position be advertised. He said he would keep me advised.

Some forty-odd years on, the true story of what happened next has only recently come to my attention. My biographer, in 1990,

In Melbourne after the return of *Wyatt Earp*, from left: Karl Oom, Stuart Campbell and Phillip Law. THE *AGE*

seeking information about my early career, examined files of the Department of External Affairs relating to the Antarctic Division in 1948. Correspondence between Campbell and the Department at that time showed how far he was prepared to go to prevent my being reappointed to *any* position in the Antarctic Division.

I left Australia again that same year for Japan, studying variations in cosmic ray intensity at sea level with latitude. Campbell wrote to External Affairs three days after I left, advising them that my appointment would terminate on 1 August. He added: 'In view of the pending re-organization of this Division, it is not proposed to recommend an extension of Mr Law's appointment.' The details are as follows.

The Head Office of the ANARE became, in July 1948, the Antarctic Division of the Department of External Affairs. Campbell's position was changed from Executive Officer of the ANARE to Officer-in-Charge of the Antarctic Division. On 12 August, a permanent Commonwealth position was advertised in the Commonwealth Gazette, Assistant-Officer-in-Charge (scientific). When I arrived back in Sydney from Japan, I found a letter from Colin Moodie, Counsellor, Department of External Affairs, drawing my attention to the advertisement. It read:

Commonwealth *Gazette*. No. 122 of 12 August, 1948
DEPARTMENT OF EXTERNAL AFFAIRS
Antarctic Division

Assistant Officer-in-Charge, Melbourne (Scientific) £648–£816*
Present actual salary—£846–1056
Applications closing—2 September 1948

Duties—Assist Officer-in-Charge in planning and organisation of scientific parties to be isolated for long periods in remote localities. Responsible in collaboration with Council for Scientific and Industrial Research for collation of all scientific knowledge gained; to act as Officer-in-Charge of the Division during his absence.

Qualifications—A University Degree in Science or Engineering together with wide experience in field work. Alternative to a University Degree—wide practical experience in Technical and Scientific development, including field work and administration, would be acceptable.

* Open to persons within and without the Commonwealth Service. The provisions of the RE-ESTABLISHMENT AND EMPLOYMENT ACT 1945 will apply in the making of appointments to this position.

I immediately lodged an application with the Department of External Affairs.

On 26 August, the Secretary of the Department of External Affairs in Canberra wrote to Campbell suggesting that three scientists—Sir Douglas Mawson, Dr Frederick White, Head of the CSIR, and Professor Leslie Martin—should join the Public Service Inspector, Mr W.H. Tucker and Campbell, in Melbourne, to constitute a selection committee.

Campbell replied, proposing that either Mr N. Warren, Head of the Commonwealth Meteorological Bureau, or Dr C.H.B. Priestley, Head of the CSIR Section of Meteorological Physics, should replace Martin. Burton agreed to this. Campbell also pointed out that it would probably be difficult for Mawson to travel to Melbourne. Copies of the various applications were sent to Tucker, Warren, Mawson, Priestley and White.

Warren said that he might be absent from Australia when the interviews took place, and sent Campbell a long letter expressing his views on what he considered would be the duties of the position and the qualifications that would be needed for an applicant to fulfil them:

DEPARTMENT OF INTERIOR
METEOROLOGICAL BUREAU
CENTRAL OFFICE

Melbourne, 7th October, 1948

Officer in Charge,
Australian National Antarctic Research Expedition,
Albert Park Barracks,
ALBERT PARK.

Dear Sir,

I am returning herewith file of applications for the position of Assistant Officer in Charge (Scientific) of the Antarctic Division after perusal.

2. In the absence of precise information as to the nature of the responsibilities and duties of this position as created, and in the projected deferment of an appointment to the position of Officer in Charge, it is somewhat difficult to determine the nature of the qualifications and experience desirable in the position.
3. Bearing in mind, however, the probability that for a period the appointee will actually be functioning in the position of Officer in Charge and combining in that position the duties of Scientific Assistant, I would conclude that selection should be with the following qualifications in view:—

    (a) A degree, or better, in Science with particular emphasis on Physics.
    (b) Experience in planning and practical administration of new or unusual ventures preferably of a scientific nature.
    (c) Personal qualifications of initiative, imagination and capacity for sympathetic administration.
    (d) If practicable, knowledge of procedures in a governmental undertaking; and,
    (e) If available, some experience in Antarctic work.

4. In his normal post of Assistant Officer in Charge (Scientific) I have assumed that his functions could be outlined as follows:—

    In association with the Planning Committee and the Executive Officers of Sciences joined in the project, determining prior to each Annual Relief,

    (i) Particular sciences to be represented at each Base during the succeeding year.
    (ii) Attending details of scientific instruments and equipment to be procured, assembled and packed for movement.
    (iii) Attending details of forms, printed matter, stationery and other supplies to be procured for each science represented.
    (iv) Securing nominations of Scientific Personnel for each relief and arranging necessary pre-selection, interviews and reports.

(v) Co-ordinating the scientific programs and objectives in the use of facilities common to all, and mutual help on Station by Scientific Personnel.
(vi) Attending to queries and requests from Station in regard to scientific matter.
(vii) After completion of each year, arranging compilation of reports and data for publication and presumably editing, printing and distribution of same where required.
(viii) Acting generally as Scientific Executive Officer in implementing the policy decisions or recommendations of the Planning Committee and as Advisory Officer to Officer in Charge in respect of the various sciences represented.
(ix) Leader of a party as and when required.

5. With these considerations in view, I am of opinion that the following applicants should be interviewed:—

M.F. Newbold
P.G. Law
G.A. Atkins
L.J. Dwyer
A.H. Praeger

6. I have, with some reluctance, excluded W/Cdr. H.W. Berry from the list. I think his personality, experience and capacity all measure up to a very high standard but his lack of Academic Scientific qualifications and experience would I think debar him for the post of Assistant Officer-in-Charge (Scientific) which forms the primary portion of the functions to be discharged.

Yours faithfully,
[signed]
DIRECTOR
(H.N. Warren)

Enc.

On 12 October, Campbell wrote to the Public Service Inspector in Melbourne stating:

> In view of the fact that it is not intended to proceed with the creation of a permanent position of Officer-in-Charge of the Division, it is probable that I will return to the Department of Civil Aviation in due course and the Assistant Officer-in-Charge will be appointed to act in that capacity.

This indicated that the government had apparently decided to move cautiously in relation to future Antarctic planning. It had already committed itself to maintaining the two sub-Antarctic stations for at least five years. Without a suitable ship, however, it was not prepared to establish a *permanent* position for the Head of the Antarctic Division.

Campbell's letter continued:

> For that reason I feel it is important that the person appointed to this position should be of a mature age and have a wide general experience with particular emphasis on handling men and departmental organisation as well as technical or scientific qualifications.
>
> Owing to the difficulty of getting Sir Douglas Mawson to Melbourne, it is proposed to interview the selected applicants and then to take the file across to him together with the Interviewing Committee's comments.

On 13 October, Campbell wrote to Mawson informing him of the date fixed for the interviewing of shortlisted applicants. He said:

> I am writing to ask if you can get over. Personally, I very much doubt it would be worth your while coming, as I know how busy you must be now and, if you can't make it, I suggest that Dr White, the Public Service Inspector, Dr C.H. Priestley and myself do the interviews, and then I bring the file across to you with our comments for your consideration, before we make a final recommendation.

Campbell added:

> With the deferment of the Antarctic program and the inevitable falling-off of scientific work other than Meteorology and perhaps Geophysics, the job will become very largely a routine one of looking after the interests of the people down there and ordering the fresh supplies through the Navy and Air Force each year.
>
> My opinion, therefore, is that in making this appointment, a wide general experience and organising ability is more important than strictly academic or scientific qualifications.

On 21 October, the Selection Committee met to interview the shortlisted applicants, including Wing-Commander Berry. Present were Tucker, Campbell, White, Priestley and Warren. Absent was Sir Douglas Mawson.

Trevor Heath had informed me of some of what was going on. As a result, I wrote, on 27 October, to Colin Moodie, Counsellor in the Department of External Affairs, Canberra, responsible for Antarctic matters, expressing my misgivings about what was happening. It was a long letter, the significant parts of which were as follows:

I am certain now, whereas before I intuitively suspected, that many apparently disconnected events are merely parts of a complete design.

That Campbell is determined that I shall not get that job I have hesitated to believe, but I cannot delude myself any longer in spite of his recent ingratiating friendliness towards me.

I shall not dwell upon the numerous humiliations I suffered during my twelve months with him. By deliberately denying me authority and executive power he made it easy for him to accuse me later of 'lack of executive ability'.

When the question of reorganisation of the office was first mooted, I asked him to inform me immediately he knew whether there was any position offering in which I might be interested. I asked him this on three separate occasions and each time he stalled me off by saying that nothing had been decided and he had no idea etc, etc.

At the same time Heath told me he had all his plans made, knew what jobs he intended advertising, etc!!

Right until the time of the advertisement he never once replied to my request.

At the interview itself the Inspector *staggered* me by asking whether I was still anxious to proceed *in view of the change in the nature of the job*! Campbell had apparently convinced him that the job was no longer a scientific one but was *purely administrative.*

I replied that I knew all along that the job would be largely administrative, that to my mind nothing had altered since I applied, and that the appointment of someone without scientific background would be a drastic blunder.

What the Public Service Inspector said or thought I do not know, but I felt he was quite favourably impressed with me at the interview.

I can recall half a dozen incidents when, with a bit of faith and enthusiasm something would have been done, but Campbell—and others—dashed cold water over everything. Remember the meeting at Canberra before the *Earp* sailed? Recall the shilly-shallying since we returned, when, with some real enthusiasm from Campbell, something would have emerged? What has been planned—scientifically—this year since I left—anything?

Look, if they appoint a 'pure administrator' even the Island programmes will degenerate into routine, unimaginative exercises in 'logistics', worth about as much as New Zealand's occupation of Campbell Island. And the Antarctic show itself will just *never* materialise. Anyway it is a *scientific* expedition, and it will be top-heavy with administrators and there won't be a senior scientist in the whole establishment!! Wouldn't some of the scientific associations (leaving out Mawson and a few others) go to town about the set-up!

Finally, what authority has Campbell to go around *altering* the *terms of appointment* as set out in the advertisement? Ask Heath to show you the letter by Warren listing what he considers to be the jobs and responsibilities of the man appointed. It is a cool, commonsense analysis of the position. Yet Campbell decides, *alone*,

> that his own interpretation (or rather modification) is to replace it!
>
> And he is smooth enough to put it over everyone but Mawson, whom he despises as a muddle-headed old 'has-been' who lives in the past! Mawson, to my mind, is one of the few with a belief in pure science and a driving faith in the ability of this country to do valuable work in Antarctica. It is these two characteristics which enabled him to conduct expeditions which have proved the most fruitful, scientifically, of any which have ever been launched.
>
> Pick a good reputable scientific bloke for this job and I'll withdraw my application. But I'll fight 'pure administrators', particularly heavy-handed ex-service administrators who believe in handling a scientific expedition like a military exercise, as long as I have any breath left.

On 3 November Campbell wrote to External Affairs stating that, in his view, neither Law nor Berry was suitable as a future OIC. Therefore the Department should:

- Defer the appointment of an Assistant Officer-in-Charge (Scientific);
- Create a position of Assistant Officer-in-Charge (Technical, Administrative);
- Ultimately, when ready to proceed with an Antarctic program, create a position of Officer-in-Charge and advertise widely in Australia and Britain. (It might or might not then be desirable to fill this vacant position.)

On 8 October, the Selection Committee met again to draft its recommendations.

Meanwhile, on 5 October, Secretary Burton had written to Dr R.V. Woolley, Commonwealth Astronomer, who had been deeply involved with Scientific Instrument and Optical Panel matters during the War, and asked him about my administrative and organisational ability. Wooley commended my work: 'He was tactful and businesslike in SIOP affairs. I should imagine that Law would be good in any company you are likely to get.'

On 11 October, Campbell sent to External Affairs the Report of the Selection Committee. It stated that 'an administrative officer with technical qualifications could carry on the routine work of servicing the island stations. Any consideration of appointing an OIC or an Assistant OIC (Scientific) should be deferred until a programme in Antarctica begins.' It went on:

> Therefore the Committee recommends:—

- that the position of Assistant Officer-in-Charge (Scientific) be not filled at present;
- that a position be created of Principal Administrative Officer (duties and qualifications accompanied the recommendation);
- when it is decided to continue with a programme of work in the Antarctic proper, consideration should be given to the appointment of an OIC of the Division and also, depending on the scope of the work envisaged, of an Assistant OIC.

The stated qualifications no longer made a degree obligatory. They were:

> Wide experience in some form of technical organisation and development, together with sound administrative experience. A university degree in Engineering or Science would be an advantage.

An External Affairs memo on the file in Canberra summarising all the above actions concludes with the significant statement:

> The general effect is that any initiative in planning the scientific side of the expeditions' work is taken away from the Division, which would have the purely routine job of finding men and getting stores.

On 18 November, Secretary Burton resolved the issue with an authoritative letter to the Secretary of the Public Service Board, Canberra. The whole of this letter is worth reproducing. It reads:

> The attached advertisement appeared in Commonwealth Gazette No. 122 dated 12th August, 1948, pursuant to approval given by the Public Service Board in its memorandum No. A47/2251 dated 18th July, 1948, for the creation of a temporary position of Assistant Officer-in-Charge (Scientific) in the Antarctic Division of this Department. At the same time, the Board determined that consideration of further recommendations of this Department for the creation of positions of Officer-in-Charge and Assistant Officer-in-Charge (Administrative) of the Division should be deferred until the Government's future Antarctic programme should be more fully worked out.
>
> A number of applications for the position were received and were forwarded to the Antarctic Division for consideration in collaboration with the Public Service Inspector, Melbourne, in accordance with established procedure. In view of the special qualifications required for the position, it was felt, however, that a panel of scientists should be constituted to consider applications in conjunction with the officer in charge of the Division, Mr S.A.C. Campbell, and the Public Service Inspector, Melbourne. It was finally decided that the scientific members of this panel would comprise Sir Douglas Mawson,

Dr F.W.G. White (Chief Executive Officer of the CSIR), Mr H.N. Warren (Director of the Commonwealth Meteorological Bureau, Department of the Interior), and Mr C. Priestly (Section of Meteorological Physics of the CSIR).

Applicants were interviewed in Melbourne on 21st October and members of the Selection Committee (excluding Sir Douglas Mawson, who transmitted a separate report to Mr Campbell) met again on 8th November to consider their recommendation. The effect of the recommendation was that the position of Assistant Officer-in-Charge (Scientific) should not be filled at present but that a position be created of Principal Administrative Officer, whose duties would be as set out in the attached list.

I forward herewith the file containing all applications, together with comments by members of the scientific panel. From this it will be seen that only two applicants, Mr P.G. Law and Wing Commander H.W. Berry, were deemed worthy of special consideration and that the latter has no scientific qualifications. In considering the applications, members of the panel apparently had in mind that, once an Assistant Officer-in-Charge (Scientific) were appointed, it would be difficult at a later stage to appoint an Officer-in-Charge over his head. They therefore did not consider the applications purely in relation to the duties of the position as advertised. This approach is erroneous, as it was at no stage contemplated by me that the Assistant Officer-in-Charge (Scientific) would have special claims to being appointed Officer-in-Charge should such a position be later created in order to carry out a full Antarctic Programme. I had proposed to make this clear to the person selected.

The main requirement at the present time is for the appointment of an individual possessing the necessary scientific ability to co-ordinate the activities of the island parties at Heard and Macquarie Islands and to put to the best possible use scientific data obtained therefrom. The routine administrative work in the Division is at present being carried out by Mr Heath, who was originally chosen as Secretary of the Antarctic Planning Committee and who had practical experience in handling the expeditions last year and at the beginning of this year, including the organisation of the Macquarie Island party during the absence of Mr Campbell. I envisage, therefore, that the functions of the Assistant Officer-in-Charge (Scientific) would be primarily scientific, coupled with general oversight and supervision of the administrative arrangements to be carried out by Mr Heath.

It will be noted from the file that Mr Warren, by his memorandum dated 7th October, rightly laid stress upon the importance of scientific qualifications and gave an expanded list of duties of the person selected. On this basis he excluded Wing Commander Berry from selection. Mr Warren's reasoning in this respect has my full endorsement, and I am satisfied, after perusal of the file, that appointment to the position of Assistant Officer-in-Charge (Scientific) should be proceeded with and that Mr Law is the most suitable applicant

> for the position. It will be seen from his application that he has in fact had administrative experience, especially during his wartime period as Assistant Secretary and Acting Secretary of the Scientific Instrument and Optical Panel of the Ministry of Munitions. Mr Law has already seen service with the Antarctic Division, having been appointed as Scientific Liaison Officer to the Planning Committee in June of 1947. Approval for this appointment was given in your A.47/2251 of 26th June, 1947. The appointment was for one year only, after which Mr Law returned to the University of Melbourne. It is therefore recommended that Mr Law be appointed to the position.
>
> As it does not appear that a Security check was carried out on Mr Law at the time of his earlier appointment, the Director, Commonwealth Investigation Service, has been requested to make the necessary enquiries and report direct to you on his findings.

On 8 December, Secretary Burton wrote to Campbell to advise him that I had been appointed Assistant Officer-in-Charge (Scientific). He concluded the letter by saying that, upon taking up the duties on 1 January 1949, I would become Acting Officer-in-Charge of the Antarctic Division. Then on 14 December, Burton wrote to the Director-General of the Department of Civil Aviation informing him that Campbell would be free to return to that Department at the end of the year, his services with External Affairs terminating on 31 December.

Finally, Burton sent a letter of appointment to me on 14 December.

One might have thought that this would be the end of the matter. However, on 15 December Campbell sent a telegram to Burton stating: '. . . I must advise most strongly against the appointment of Assistant Scientific Officer.' External Affairs merely responded that all factors had been taken into account, and that they were proceeding with the appointment.

On 16 December Campbell reiterated his strong opinions, concluding: '. . . The position should be re-advertised or else the recommendations of the Selection Committee should be adopted.' On 20 December, he sent an urgent teleprinter to External Affairs stating: '. . . Other members of the Committee contacted feel this is an inadvisable step.' On 21 December, he wrote to External Affairs proposing that he should stay on for a further two months, 'because neither Law nor anyone else could take effective control at this stage'.

The Departmental reply was that they would ask Civil Aviation to allow him to devote a portion of his time to guiding me. Any information or advice he could give would be greatly appreciated.

After 1 January, I did not see Campbell again.

He had one bullet left to fire, however. He persuaded White to write to External Affairs, which White did, on 4 January 1949. White stated that Campbell had shown him copies of the correspondence regarding the appointment, and that he supported the decision of the Selection Committee. He thought that the sudden handing over to me in the middle of the busy expedition season might cause difficulties.

Burton replied on 7 January, explaining, as he had done to the Public Service Board, the erroneous approach of the Selection Committee to the duties of the position as advertised. White wrote back on the 11th saying: 'I now appreciate your point of view. Law will be a very good man for the position of Assistant Officer-in-Charge (Scientific).'

I should mention that, from that time onwards, I had no greater supporter in all that I did than Frederick White.

I should also mention that the reluctance of the Department to appoint an Officer-in-Charge continued. I remained Acting Officer-in-Charge for the next three years, despite my demonstrated ability to handle the job and the growing appreciation that there would be no turning back from the ultimate objective of establishing a permanent scientific station on the Antarctic Continent.

POSTSCRIPT

Back at the Physics Department at Melbourne University after my subsequent trip to Japan across the tropical zone, Mac and I worked on our cosmic ray records. Then, one night in September, the hut annexe in which we were housed burnt to the ground. I was shocked: the scientific results of two voyages destroyed! I could hardly credit the news when I arrived at work the next morning.

Mac and I wandered disconsolately through the blackened skeleton of the hut and examined the charred remains of the wooden filing cabinet that had held our paper records. The sheets on which we had daily entered the results were of standard size, ruled into columns, on which were listed the counting rates of the cosmic ray recorders each hour, together with meteorological and other data. These sheets had been clamped tightly into a number of separate hard-backed folders.

Among the ruins of the hut we found what was left of our records. To our amazement, the hard-packed pages, although soaked with water from the firemen's hoses, and charred to a depth of half an inch all around the edges, were otherwise unharmed. The first few pages at the front and the last few at the end were burnt off, as also were the covers, but the entries, luckily, had been made with the recently invented ballpoint pens whose ink was waterproof, and therefore had not run when wet.

We carried blackened stacks of saturated paper to a deserted practical laboratory of the Department. Then, with great care, we peeled each page off and spread it to dry on the floor. We covered the whole lab! Next day we used scissors to cut off a half-inch of blackened border from each page. Then we were able to put the sheets back in order again, and resume our analysis of them.

From the results, we published two papers, one on each voyage, in the *Australian Journal of Physics*.

To cap 1948 I found myself, in December, appointed Acting Officer-in-Charge of the Antarctic Division of the Department of External Affairs, with duty commencing on 1 January 1949.

APPENDIX 1
Ship's Officers, crew and supernumeraries

SHIP'S OFFICERS

Captain	Commander Karl Eric Oom, OBE, RAN
First Officer	Lieut.-Crd. William Frank Cook, RAN
Second Officer	Lieut. Norman Harold Stephen White, RAN Navigator
Third Officer	Lieut. John Gestwyck Yule, RAN
Chief Engineer	Temporary Engineer Lieut.-Crd. Harold Frederick Irwin, RANR(S)
Administrative Officer	Acting Commissioned Boatswain John William Homewood, RAN

CREW

Brand, R.B.	Engine Room Artificer
Clarke, J.P.	Officers' Steward
Doody, K.J.	Petty Officer—Chief Bosun's Mate
Flack, H.J.	Engine Room Artificer
Fulton, F.J.	Able Seaman
Gilbert, F.A.	Leading Seaman
Hoy, W.L.	Chief Engine Room Artificer
Holmes, R.L.	Able Seaman
Liddell, T.W.	Chief Petty Officer, Wireless Telegraphist

Manser, R.J.	Stoker
Parfrey, S.J.	Cook
Rosser, N.D.	Leading Stoker
Rodgers, C.T.	Able Seaman
Roberts, K.J.	Stores Assistant
Smith, J.K.	Acting Petty Officer, Seaman
Sutherland, N.	Chief Shipwright
Skinner, G.F.	Leading Wireless Telegraphist
Swinnerton, E.C.	Able Seaman
Tame, N.W.	Able Seaman
West, C.V.	Petty Officer, Stoker
Walsh, A.G.	Leading Stoker
Wallace, W.R.	Able Seaman, Senior
Young, R.C.	Able Seamen

Supernumeraries

Group-Capt. Stuart Campbell	OIC, ANARE (second voyage only)
Phillip Law	Chief Scientist, Physicist
Dr Fritz Loewe	Meteorologist
Edward McCarthy	Geophysicist
Dr Tony Bond	Medical Officer
Sqd.-Ldr Robin Gray, RAAF	Pilot
Flt.-Sgt 'Spike' Jones, RAAF	Mechanic
Laurie Le Guay	Photographer

Appendix 2
The Beaufort Scale of wind strengths

Beaufort number	Average wind velocity (knots)	Seaman's description of wind	Steel full-rigged ship, deep loaded
0	0	Calm	—
1	2	Light air	Steerage way only
2	5	Light breeze	2 knots in full sail
3	9	Gentle breeze	3–4 knots in full sail
4	14	Moderate breeze	5–6 knots in full sail
5	19	Fresh breeze	Can just carry royals and light staysails
6	24	Strong breeze	
7	30	Moderate (half) gale	
8	37	Fresh gale	
9	44	Strong gale	
10	52	Heavy (whole) gale	Main topsail only
11	60	Storm	Storm staysail or trysail only
12	Over 80	Hurricane	No canvas can stand

APPENDIX 3
Glossary of ice terms

Bay ice: Fast ice of more than one winter's growth, which may be nourished also by surface layers of snow. Thickness of ice and snow up to about 2 m above sea-level. When bay ice becomes thicker than this it is called an ice shelf.

Bergy bit: A massive piece of sea ice or disrupted hummocked ice; also medium-sized piece of floating glacier ice. Generally less than 5 m above sea-level, and not more than about 10 m across.

Beset: Situation of a vessel surrounded by ice and unable to move.

Brash ice: Small fragments of floating ice, not more than 2 m across; the wreckage of other forms of ice.

Close pack ice: Composed of floes mostly in contact, such that navigation is difficult even for specially constructed vessels and ice-breaker assistance may be required.

Crack: Any fracture or rift in sea ice not sufficiently wide to be described as a lead. It is usually possible to jump across a crack.

Fast ice: Sea ice of greatly varying width which remains fast along the coast, where it is attached to the shore, to an ice wall, to an ice front, or over shoals, generally in the position where originally formed. Fast ice may extend 400 km from the coast.

Floe: A piece of sea ice other than fast ice, large or small. Light floes are anything up to a metre in thickness.

Frazil ice: Fine spicules or plates of ice in suspension in water. The first stage of freezing, giving an oily or opaque appearance to the surface of the water.

Glacier: A mass of snow and ice continuously moving from higher to lower ground or, if afloat, continuously spreading.

Glacier tongue: An extension of a glacier, projecting seaward, and usually afloat.

Growler: A piece of ice almost awash, smaller than a bergy bit.

Hummocked ice: Sea ice piled haphazardly one piece over another, and which may be weathered.

Hummocking: Process of pressure formation by which level ice becomes broken up into humps and ridges. When the floes rotate in the process it is termed 'screwing'.

Iceberg: A large mass of floating or stranded ice broken away from a glacier or from an ice shelf. Often of considerable height (in any case more than 5 m above the level of the sea).

Ice blink: Yellowish white luminous glare in the sky produced by the reflection on the clouds of pack-ice or an ice sheet which may be beyond the range of vision.

Ice cover: The amount of sea ice encountered; measured in eighths of the visible surface of the sea covered with ice.

Ice edge: The boundary at any given time between the open sea and sea ice of any kind, whether drifting or fast. When wind or swell compact the ice edge it may be called a compacted ice edge; when dispersed, it may be called an open ice edge.

Ice rind: A thin, elastic, shining crust of ice, formed by the freezing of sludge on a quiet sea surface. Thickness less than 5 cm.

Ice sheet: A continuous mass of ice and snow of considerable thickness and large area.

Ice shelf: A floating ice sheet of considerable thickness. Ice shelves are usually of great horizontal extent and have a level or gently undulating surface.

Lead: A navigable passage through pack ice; a lead may still be so called even if covered with young ice.

Open pack ice: Composed of floes seldom in contact and with many leads and pools.

Pack ice: Any area of sea ice other than fast ice, no matter what form it takes or how it is disposed.

Pancake ice: Pieces of newly formed sea ice usually approximately circular in shape with raised rims, and from one-third to 2 m in diameter.

Polynya: A large area of open water, surrounded by sea ice and found in the same region every year. One side is sometimes formed by the coast. Polynyas commonly occur off the mouths of big rivers.

Pressure ice: A general term for sea ice which has been squeezed, and in places forced upwards when it can be described as rafted ice, hummocked ice or pressure ridge.

Pressure ridge: A ridge or wall of hummocked ice where one floe has been pressed against another.

Rafting: The overriding of one floe on another, the mildest form of pressure. Hence 'rafted ice'.

Rotten ice: Sea ice which has become honeycombed in the course of melting, and which is in an advanced state of disintegration.

Sludge: A later stage of freezing than frazil ice, when the spicules and plates of ice coagulate to form a thick soupy layer on the surface. Sludge reflects little light, giving the sea a matt appearance.

Tabular berg: A flat-topped iceberg. Most tabular bergs form by breaking from an ice shelf and show horizontal firm snow layers.

Tide crack: The fissure at the line of junction between an immovable icefoot or ice wall and fast ice, the latter being subject to the rise and fall of the time.

Tongue: A projection of the ice edge up to several kilometres in length, caused by wind and current.

Winter ice: More or less unbroken, level sea ice of not more than one winter's growth originating from young ice. Thickness from 15 cm to 2 m.

Young ice: Newly formed level ice in the transition stage of development from ice rind or pancake ice to winter ice. Thickness from 5 to 15 cm.

Source: Terence Armstrong and Brian Roberts, *Illustrated Ice Glossary*, Scott Polar Research Institute, Cambridge, 1956.

INDEX

Page numbers in *italics* indicate illustrations